The Common Objects of the Country

J. G. Wood

Alpha Editions

This edition published in 2021

ISBN : 9789355756121

Design and Setting By
Alpha Editions
www.alphaedis.com
Email – info@alphaedis.com

As per information held with us this book is in Public Domain.
This book is a reproduction of an important historical work. Alpha Editions uses the best technology to reproduce historical work in the same manner it was first published to preserve its original nature. Any marks or number seen are left intentionally to preserve its true form.

PREFACE

In the following pages will be found short and simple descriptions of some of the numerous objects that are to be found in our fields, woods, and waters.

As this little work is not intended for scientific readers, but simply as a guide to those who are desirous of learning something of natural objects, scientific language has been studiously avoided, and scientific names have been only given in cases where no popular name can be found. In so small a compass but little can be done; and therefore I have been content to take certain typical objects, which will serve as guides, and to omit mention of those which can be placed under the same head.

Every object described by the pen is illustrated by the pencil, in order to aid the reader in his researches; and the subjects have been so chosen that no one with observant eyes can walk in the fields for half-an-hour without finding very many of the objects described in the book.

CHAPTER I

EYES AND NO EYES—DIFFICULTIES OF OBSERVERS—THE BATS—LONG-EARED BAT—ITS UTILITY—SPORT AND MURDER—SONG OF THE BAT—A BRAVE PRISONER—HOW BATS FEED—HAIR OF BAT AND MOUSE—WING OF THE BAT—THE FIELD-MOUSE—ITS STEALTHY MOVEMENTS—HARVEST MOUSE—WATER RAT—AN INNOCENT VICTIM.

Every one has read, or at least heard of, the tale entitled "Eyes and no Eyes"; which tale is to be found in the *Evenings at Home*. Now this story, or rather the moral of it, is, in my opinion, as often used unfairly as rightly.

Although there are those who pass through life with closed eyes and stopped ears, yet there are many more who would be glad to use their eyes and ears, but know not how to do so for want of proper teaching. To one who has not learned to read, the Bible itself is but a series of senseless black marks; and similarly, the unwritten Word that lies around, below, and above us, is unmeaning to those who cannot read it.

Many would like to read, but cannot do so; and it is in order to help such, to bring before them the first alphabetical teaching, that the following pages are written.

It is no matter of marvel that many an observant person becomes bewildered among natural objects; that he is lost amid the variety of animal, vegetable, and mineral life in which he lives; and that, after vainly attempting to comprehend some simple object, he finds himself baffled, and so in despair ceases to inquire into particulars, and contents himself with admiration of and love for nature in general.

Objects change so rapidly and so constantly, that there is hardly time to note a few remarks before the season has passed away; the object under examination has changed with it, and a year must elapse before that investigation can be continued.

From experience I know how valuable are even a few hints by which the mind can be directed in a straight course without wasting its strength and losing its time by devious wanderings. Only hints can be given, for the limits of the volume forbid any lengthened discussion of single objects; and, besides, the mind is more pleased to work out a subject according to its own individuality than to have it laid down as completed, and to be forbidden to go any further.

Almost every object that is described by the pen will be figured by the pencil, in order to assist the reader in identifying the creature in an easier manner than if it were merely described in words.

Of the birds I shall not be able to treat, as they alone would occupy the entire space of this volume; and, for the same reason, only a short account can be given of each object.

As in the scale of creation the mammals fill the highest place, we will speak of them first, taking, as far as possible, each creature in its own order.

Perhaps there are few people who would not feel some surprise when they learn that the very highest of our British animals is the Bat. Usually the bat is looked upon with rather a feeling of dread, and is regarded as a creature of such ill-omen that its very presence causes a shudder, and its approach would put to flight many a human being.

There is certainly some ground for this feeling; for the night-loving propensities of the creature, its weird-like aspect, its strange devious flight, and more especially its organs of flight, are so interwoven with the popular ideas of evil and its ministers, that bats and imps appear to be synonymous terms.

Painters always represent their imps as upborne by bats' wings, furnished with several supplementary hooks; and sculptors follow the same principle.

In consequence, all bats and objects connected with bats are viewed with great horror, with two exceptions: a cricket-bat and a bat's-wing gas-burner.

Now, I cannot but think that this is very hard on the bats. It is said that the African negroes depict and describe *their* evil spirits as white; and that, in consequence, the negro children fly in consternation if perchance a white man comes into their territory.

Yet, a white man is not so very horrid an object after all, if one only dare look at him; and the same remark holds good with the bats.

COMMON LONG-EARED BAT.

A very pretty creature is a bat, more especially the long-eared species, *Plecotus communis*, as it is scientifically called, and its habits are most curious. It is well worth the time to watch these little creatures on a warm summer's night, as they flit about in the air, and to note the enjoyment of their aërial hunt. They are fearless animals; and provided that the observer remains tolerably still and does not speak, bats will often flit so close to his face that he could almost catch them in his hand.

Their flight is very singular, and reminds one of the butterfly in its apparently vague flitting. Indeed, there are many large moths that fly by night who can hardly be distinguished from the bats, if the evening be rather dark, so similar are they in their mode of journeying through the air.

From this peculiarity of flight, they are accounted difficult marks for a gun; and it is unfortunately a custom with some ruthless powder-burners to practise by day at swallows and by night at bats. Now, even putting the matter in its lowest form, it is wrong to shoot swallows; for they are most useful birds, and serve to thin the host of flies and other insects that people the summer air.

As regards the swallow, this is well known, and does serve to protect it from some persons who have more compassion than the generality. Moreover, the swallows, swifts, and martins are extremely pretty birds, and their beauty is in some degree their shield.

But the bat is as useful a creature as the swallow, and in the very same way; for, when the evening comes on, and the swallow retires to its nest, the bat issues from its home and takes up the work just where the swallow leaves it—the two creatures dividing the day and night between them. Therefore, let those who refrain from swallow shooting include the bat in their free list.

Some there are whom nothing can restrain from killing, for the instinct of slaughter is strong in them. With them nothing is valuable unless it is to be killed. If it can be eaten afterwards, so much the better; but the great enjoyment consists in the mere act of killing.

They contrive to disguise the ugliness of the thing by giving it any name but the right one; but, in spite of the name, the thing exists. And I wonder, if they were to look very closely into themselves, whether they would not find there a decided desire to kill men, provided that they had no reason to dread the consequences. Those who have practised the sport unanimously say that nothing is so exciting as man-hunting and killing and that all other sport is tame in comparison.

The chief name under which this profanity is disguised is that of "Sport," a word which always reminds me of the "Frog and Boys" fable. There are actually men who are audacious enough to declare that there is no cruelty in "sport"; that foxes are charmed at being hunted, and that pheasants derive a singular gratification from getting shot. Now, I never was either a fox or a pheasant; but I entirely repudiate the assertion that any animal likes to be chased or to be wounded; and, moreover, I disbelieve the sincerity of the man who can say such a thing. If he says openly that he finds excitement in the chase, and means to gratify himself without any reference to the feelings of the creatures which he chases, I can understand while I disapprove. But when a man justifies himself by asserting that any animal likes to be hunted, I can hardly find epithets too contemptuous for him; and I could see him run the gauntlet among the Sioux Indians with but small pangs of conscience.

Some again call themselves Naturalists, and under the shelter of that high-sounding name occupy themselves in destroying nature. The true naturalist never destroys life without good cause, and when he does so, it is with reluctance, and in the most merciful way; for the life is really the nature, and that gone, the chief interest of the creature is gone too. We should form but a poor notion of the human being were we only to see it presented to our eyes in the mummy; and equally insufficient is the idea that can be formed of an animal from the inspection of its outward frame. Nature and life belong to each other; and, if torn asunder, the one is objectless and the other gone.

Lastly, let me remind those who find such gratification in destroying, that the word "Destroyer" is in the Greek language "Apollyon".

As we do not intend to treat of the dead and dried bodies of animals, but of their active life, we return to our bat flitting in the evening dusk, and, instead of shooting him, watch his proceedings.

Every creature is made for happiness, and receives happiness according to its capacity; and it is very wrong to suppose that, because *we* should be miserable if we led the life of a vulture, or a sloth, or a bat, therefore those creatures are miserable. In truth, the vulture is attracted to, and feels its greatest gratification in, those substances which would drive us away with averted eyes and stopped nostrils. The sloth is, on the authority of Waterton, quite a jovial beast, and anything but slothful when in his proper place; and as for the bat, it sings for very joy. True, the song is not very melodious, neither is that of the swift, or the peacock, nor, perhaps, that of the Cochin-China fowl, but it is nevertheless a song from the abundance of the heart.

There are many human ears that are absolutely incapable of perceiving the cry of the bat, so keen and sharp is the note; a very razor's-edge of sound.

More than once I have been standing in a field over which bats were flying in multitudes, filling the air almost oppressively with their sharp needle-like cries. Yet my companion, who was a musician, theoretically and practically, was unable to hear a sound, and could not for some time believe me when I spoke of the noisy little creatures above.

The sound bears some resemblance to that produced by a slate-pencil when held perpendicularly in writing on the slate, only the bat's cry is several octaves more acute. I never but once heard the sound correctly imitated, and that was done by a graceless urchin, during a long sermon one Sunday morning. He had contrived to arrange two keys in such a manner that, when grated over each other, they produced a squeaking sound that exactly resembled the cry uttered by the bat. So, by judicious management of his keys, he kept the congregation on the look-out for the bat, and beguiled the time much to his satisfaction.

Of so piercing and peculiar a nature is the cry, that it gives no clue to the position or distance of the creature that utters it, and it seems to proceed indiscriminately from any portion of the air towards which the attention happens to be directed. The note of the grasshopper lark possesses somewhat of the same quality.

Even in confinement the bat is an interesting creature, and discovers certain traits of character and peculiarities of habit which in its wild state cannot be seen. I might here refer to several stories of domesticated and tamed bats; but as they have already been given to the world, and my space is limited, I prefer to give my own experiences.

Not long ago, I received a message from a neighbouring grocer, requesting me to capture a bat which had flown into the shop, and which no one dared touch.

When I arrived, the creature had taken refuge on an upper shelf, and had crawled among a pile of sugar-loaves that were lying on their sides after the usual custom. We pulled out several loaves near the spot where the bat was last seen, and by casting a strong light from a bull's-eye lantern, discovered a little black object snugly ensconced at the very back of the shelf.

I pushed my hand towards the spot, but for some time could not seize the creature, as it was so tightly packed, and squeezed into a corner. At last the bat gave a flap with one of the wings, which I caught, and so gently drew my prisoner forwards.

He was a brave little fellow, as well as discreet, and bit savagely at my fingers. However, his little tiny teeth could not do much damage, and I put him into a cage which I brought with me.

The cage was originally made for the reception of mice, and was of a rude character—the back and ends being of wood and the front of wire. In a very few minutes after his entrance into the cage, the bat climbed up the wooden back, by hitching his claws into the slight inequalities of the wood, and there hung suspended, head downwards.

When so placed, his aspect was curious enough. The claws of the hind legs being fixed into a crevice, so as to bear the weight of the body, the wings were then extended to their utmost, and suddenly wrapped round the body. At the same time the large ears were folded back under the wings and protected by them, the orifice of the ear itself being guarded in a very singular manner.

If the reader will refer to the figure of the bat on page 4, he will see that inside the great ear is a sharply-pointed membrane, somewhat resembling a second ear. This membrane is called the "*tragus*," and when the large ears are tucked away out of sight, the tragus remains exposed, and gives the creature a very strange appearance.

When the bat is living, the ears are of singular beauty. Their substance is delicate, and semi-transparent if viewed against the light; so much so, indeed, that by the aid of a microscope the circulation of the blood can be detected. As the creature moves about, the ears are continually in motion, being thrown into graceful and ever-changing curves. If people only knew what a pretty pet the long-eared bat can become, they would soon banish dormice and similar creatures in favour of bats.

It was rather a remarkable circumstance, that the bat of which I have just been speaking would not touch a fly, although one which I had in my possession some ten years since would eat flies and other insects readily. Whenever it took the insect, it daintily ate up the abdomen and thorax, rejecting the head, wings, and legs. But my second bat entirely refused insects of any kind, and would eat nothing but raw beef cut up into very small morsels. I never had a pet so difficult to feed.

If the meat were not perfectly fresh, or if it were not cut small enough, the bat would hardly look at it. Now if a bit of raw meat about the size of a large pin's head be placed in the air, a few minutes will dry and harden its exterior; and when this was the case, my bat did not even notice it. So I had to make twenty or more attempts daily before the creature would condescend to take any food.

When, however, it *did* eat, its mode of so doing was remarkable enough. It seized the meat with a sharp snap, retreated to the middle of the cage, sat upright—as in the engraving already alluded to—thrust its wings forward to form a kind of tent, and then, lowering its head under its wings, disposed of the meat unseen.

From the movement of the neck and upper portion of the head, it would be seen that the creature ate the meat much after the manner of a cat; that is, by a series of snaps or pecks; for the teeth are all sharply pointed, and have no power of grinding the food. These teeth can be seen in the accompanying sketch of a bat's skull.

In many parts of England the bats are called "Flitter-mice," and are thought to be simply mice plus wings. This opinion has been formed from the resemblance between the general shape, and especially that of the fur, of the two animals. But if we look at the teeth, we find at once that those of the bat are sharp and pointed, extending tolerably equally all round the jaw-bone; while the teeth of the mice are of that chisel-shaped character found in the rabbit and other rodent animals.

Now if we turn to the fur, and examine it with a microscope, we shall there find characteristics as decided as those of the teeth.

On this page is the magnified image of a single hair, taken from the long-eared bat. It will be seen that the outline of the hair is deeply cut, and the markings run in a double line. These markings and outlines are caused by the structure of the hair, which is covered with a regular series of scales adhering but loosely to its exterior. These scales can be removed by rough handling, and therefore the aspect of the hair can be much altered.

Let us now take a hair from the common mouse, and place it under the microscope. This being done, we find the result to be as shown in the accompanying cut.

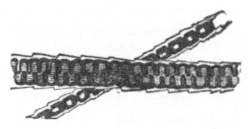

The two objects here shown are two portions of the same hair; the upper one showing the middle of the hair, and the lower being taken from a portion nearer the root. Both these specimens were taken by myself from the animals, and drawn by myself by means of the Camera Lucida, so that they are to be depended on.

To return to my caged bat.

Although it did not do much in the eating way, it frequently came to the water vessel and drank therefrom; but it was so timid when drinking, that I could not see whether it lapped or drank. When disturbed, it used to scuttle away over the floor, in a most absurd manner, but with some speed. Sometimes it tried to drink by crawling to a spot just over the vessel, and lowering itself until its nose was within reach of the water; but the distance was too great for the attempt to be successful. In its wild state, the bat hunts insects, as they hover over the surface of water, and drinks as it flies, by dipping its head in the water while on the wing.

I rather think that my bat must have received some injury from the brooms and caps that were aimed at it when it entered the shop, for it only lived a fortnight or so, and one morning I found it hanging by its hind claws from the roof of the cage, quite dead.

I believe that bats generally die while thus suspended, for it is a very common thing to find plenty of suspended bats, dry and mummified, when entrance is made into an unfrequented cave, or a hollow tree cut down, or, indeed, when any bat-haunted spot is examined.

In speaking of the bat, I have used popular terms, and therefore have employed the word "wing". But the apparatus of the bat is not a wing at all, but only a developed hand. Let the reader spread his hand as wide as he can, and he will see that between each finger, and especially between the forefinger and the thumb, the skin forms a kind of webbing, something of the same kind as that on the feet of ducks and other aquatic birds.

Now if the bones of the fingers were drawn out like wire until they became some seven or eight feet long, and the skin between them were extended to the nails of the elongated fingers, we should have a structure analogous to that of the bat's wing. The thumb joint is left comparatively free; and by means of this joint, and the hooked claw at its extremity, the creature walks on a level surface, or can crawl suspended from a beam or a trunk. It is very curious to see the bat stretching out its wings and feeling about for a convenient spot whereon to fix the hooks.

So tenacious are these hooks, that the baby bat is often found enjoying an airing by clinging to the body of its mother, and holding firm, while she flies in search of prey.

It is true that the little creature is suspended with its head downwards; but it appears quite comfortable, nevertheless. Bat-children do not suffer from determination of the blood to the brain. Neither do certain human children, it seems, if we are to take as a criterion those whom we see hanging half out of perambulators, fast asleep, and rolling from side to side with every movement of the vehicle.

Both my bats were very particular, not to say finicking, about their personal appearance. They bestowed much time and pains on the combing of their fur, and specially seemed to value a straight parting down the back.

It was most interesting to watch the little thing parting its hair. The claw was drawn in a line straight from the top of the head to the very tail, and the fur parted at each side with a dexterity worthy of an accomplished lady's-maid. The same habit has been observed in other bats that have been tamed.

There are more than twenty British bats, but the habits of all are very similar; and so I prefer to take the prettiest, and, having described it, to leave the remaining species for a future occasion.

Pass we now from the Flitter-mouse to the Mouse.

In the fields, in the farm-yards, in the barns, and in the ricks are to be found myriads of certain little animals called Field-mice. Acting on the principle that I have just laid down, I shall take the most common and I think the prettiest species—the Common Short-tailed Field-mouse, represented on next page.

The fur of this creature is strongly tinged with red, and by its colour alone it is easily to be distinguished from the common grey or brown mouse. Its tail is short and stumpy, looking as if it had suffered amputation at an early period of life, and its nose is more rounded than that of the common mouse. Indeed, it has a very bluff and farmer-like aspect, and looks as if it ought to wear top-boots.

SHORT-TAILED FIELD-MOUSE.

Common as these little creatures are, they are seldom seen, because they keep themselves so close to the ground, and assimilate so nearly with it in colour, that they cannot easily be descried among the grass stalks, under shelter of which they pursue their noiseless way.

Their speed is not nearly so great as that of the house-mice, but they are much more difficult to catch; for they wind among the grass so lithely, and press upon the earth so closely, that the fingers cannot readily close on them, even when they are discovered.

From this facility of avoiding observation and capture, they seem to derive much audacity, and run about a field in fear of nothing but the kestrel.

When first I made a personal acquaintance with these creatures, it was under rather peculiar circumstances. There is a certain field, which was given up to football, cricket, hockey, and similar games, as soon as the grass was converted into hay and removed. One day I was very tired with running, and lay down to rest on a pile of coats that had been laid aside; my eyes were fixed on one spot of earth, just visible between the grass stalks, but without any particular object. Presently I thought I saw a something red glide across the spot, but was not certain. However, I leaned over the place and a little farther on saw the same thing again. So I made a sharp pounce at the object, and found that I had caught a short-tailed field-mouse.

Now here was this impertinent little animal taking a walk close to the wicket, in spite of the bats, ball, and runners. In order to watch its proceedings, I released it, and followed it in its progress. After watching for a few minutes, I happened to look up for a moment; and when I again looked for the creature, it was gone, and I could not find it again.

Subsequently I became sufficiently expert to find them whenever I wished; and if I wanted a field-mouse, seldom had to examine more than a square yard of ground without finding one.

They are very injurious little creatures, for they are not content with eating corn, but nibble the young shoots of various plants, and sometimes strip young trees of their bark.

Fortunately we have allies in air and on earth, in the persons of owls and kestrels, stoats and weasels, or the damage done by these red-skinned marauders would be more than serious.

Some idea of the damage that may be done by the aggregate numbers of these small quadrupeds may be formed from the fact that in Dean Forest and the New Forest great numbers of holly plants were entirely destroyed by them, they having eaten off the bark for a distance of several inches from the ground. And other trees were favoured with the notice of the field-mouse, but in a different mode. Great numbers of oak and chestnuts were found dead, and pulled up; and when pulled up, it was seen that their roots had been gnawed through, about two inches below the level of the ground.

Various modes of destroying the marauders were put in practice, such as traps, poison, &c., but the most effectual was, as effectual things generally are, the most simple.

A great number of holes were dug in the ground, about two feet long, eighteen inches wide, and eighteen inches deep. This is the measurement at the bottom of the hole; but at the top the hole was only eighteen inches long and nine wide, so that when mice fell into it, they were unable to escape.

In these holes upwards of forty thousand mice were taken in less than three months, irrespective of those that were removed from the holes by the stoats, weasels, crows, magpies, owls, and other creatures.

Like most of the mouse family, the field-mouse is easily tamed; and I have seen one that would come to the side of its cage, and take a grain of corn from its owner's fingers.

HARVEST-MOUSE.

There is another kind of mouse which may be found in the autumn, together with its most curious nest. This is the Harvest-mouse, the tiniest of British quadrupeds, two harvest-mice being hardly equal in weight to a halfpenny.

The chief point of interest in this little creature is its nest, which is not unfrequently found by mowers and haymakers when they choose to exert their eyes.

One of these nests, that was brought to me by a mower, was about the size of a cricket ball, and almost as spherical. It was composed of dried grass-stems, interwoven with each other in a manner equally ingenious and perplexing. It was hollow, without even a vestige of an entrance; and the substance was so thin that every object would be visible through the walls. How it was made to retain its spherical form, and how the mice were to find ingress and egress, I could not even imagine. The nest was fastened to two strong and coarse stems of grass that had grown near a ditch, and had overgrown themselves in consequence of a superabundance of nourishment.

WATER-RAT.

If we walk along the bank of a stream or a pond, we shall probably hear a splash, and looking in its direction, may see a creature diving or swimming, which creature we call a Water-rat; to the title of Rat, however, it has but little right, and ought properly to be called the "Water-vole".

On examining the banks we shall find the entrance to its domicile, being a hole in the earth, just above the water, and generally, where possible, made just under a root or a large stone. Sometimes the hole is made at some height above the water, and then it often happens that the kingfisher takes possession, and there makes its home. Whether it ejects the rat or not I cannot say, but I should think that it is quite capable of doing so. Many a time I have seen the entrance to a rat-hole decorated with a few stray fish-bones, which the rustics told me were the relics of fish brought there and eaten by the water-rat. But I soon found out that fish-bones were a sign of kingfishers, and not of rats; and so guided, found plenty of the beautiful eggs of this beautiful bird. Excepting the eggs of swallows and martins, I hardly know any so delicately beautiful as those of the kingfisher, with their slight rose tint and semi-transparent shell. But, alas! when the interior of the egg is removed, the pearly pinkiness vanishes, and the shell becomes of a pure white, very pretty, but not containing a tithe of its former beauty.

The piscatorial propensities of the kingfisher are not the only cause of the slanderous reports concerning the water-vole, and its crime of killing and eating fish. The common house-rat often frequents the water-side; and, it being a great flesh-eater, certainly does catch and eat the fish.

But the water-rat is a vegetable feeder, and I believe almost, if not entirely, a vegetarian in diet. That it is so in individual cases, at all events, I can personally testify, having seen the creature engaged in eating.

In former days, when I thought the water-rats ate fish, I waged war against them, for which warfare there are great facilities at Oxford. However, a circumstance occurred which showed me that I had been wrong.

I saw a water-rat sitting on a kind of raft that had formed from a bundle of reeds which had been cut and were floating down the river. Seeing it busily at work feeding, I took it for granted that it was eating a captured fish, and shot it accordingly, stretching it dead on its reed raft.

On rowing up to the spot, I was rather surprised to find that there was no fish there; and on examining the reeds, I rather wondered at the regular grooves cut by my shot. But a closer inspection revealed a very different state of things; namely, that the poor dead rat was quite innocent of fish eating, and had been gnawing the green bark from the reeds, the grooves being the marks left by its teeth. After this I gave up rat shooting on principle.

Once, though, a rather curious circumstance occurred.

In my possession was a pet pistol, which would throw a ball with great accuracy, and I considered myself sure of an apple at sixteen paces. One day, just as I was standing by a branch of the river Cherwell, I saw a water-rat sitting on the root of a tree at the opposite side of the river, and watching me closely. The river was not above twelve or fourteen yards wide; and the rat presented so good a mark that I fired at him, and, of course, expected to see him on his back.

But there sat the rat, quite still on the stump, and about two inches below him the round hole where the bullet had struck.

As the creature seemed determined to stay there, I reloaded, and took a good aim, determined to make sure of him. As the smoke cleared away, I had the satisfaction of seeing the rat in exactly the same position, and another bullet-hole close by the former. Four shots I made at that provoking animal, and four bullets did I deposit just under him. As I was reloading for a fifth shot, the rat walked calmly down the stump, slid into the water, and departed.

Now, whether he acted from sheer impertinence, or whether he was stunned by the violent blow beneath him, I cannot say. The latter may perhaps be the case, for squirrels are killed in North America by the shock of the bullet against the bough on which they sit, so that no hole is made in their skins, and the fur receives no damage. Perhaps the rat was actuated by a supreme contempt for me and my shooting powers; and, as the result showed, was quite justified in his opinion.

CHAPTER II.

SHREW-MOUSE—DERIVATION OF ITS NAME—SHREW-ASH—THE SPIRIT AND THE LIFE—WATER-SHREW—ITS HABITS—THE MOLE—MOLE-HILL—A PET MOLE—THE WEASEL.

I have already mentioned that the water-rat has little claim to the title of rat; and there is another creature which has even less claim to the title of mouse. This is the Shrew, or Shrew-mouse as it is generally called. This creature bears a very close relationship to the hedgehog, and is a distant connection of the mole; but with the mouse it has nothing to do.

SHREW-MOUSE.

Numbers of the shrews may be found towards the end of the autumn lying dead on the ground, from some cause at present not perfectly ascertained. If one of these dead shrews be taken, and its little mouth opened, an array of sharply-pointed teeth will be seen, something like those of the mole, very like those of the hedgehog; but not at all resembling those of the mouse.

The shrew is an insect and worm-devouring creature, for which purpose its jaws, teeth, and whole structure are framed. A rather powerful scent is diffused from the shrew; and probably on that account cats will not eat a shrew, though they will kill it eagerly.

On examining Webster's Dictionary for the meaning of the word "shrew," we find three things.

Firstly, that it signifies "a peevish, brawling, turbulent, vexatious woman".

Secondly, that it signifies "a shrew-mouse".

Thirdly, that it is derived from a Saxon word, "*screawa*," a combination of letters which defies any attempt at pronunciation, except perhaps by a Russian or a Welshman.

Now, it may be a matter of wonder that the same word should be used to represent the very unpleasant female above-mentioned, and also such a pretty, harmless little creature as the shrew. The reason is shortly as follows.

In days not long gone by, the shrew was considered a most poisonous creature, as may be seen in the works of many authors. In the time of Katherine—the shrew most celebrated of all shrews—any cow or horse that was attacked with cramp, or indeed with any sudden disease, was supposed to have suffered in consequence of a shrew running over the injured part. In those days homœopathic remedies were generally resorted to; and nothing but a shrew-infected plant could cure a shrew-infected animal. And the shrew-ash, as the remedial plant was called, was prepared in the following manner.

In the stem of an ash-tree a hole was bored; into the hole a poor shrew was thrust alive, and the orifice immediately closed with a wooden plug. The animal strength of the shrew passed by absorption into the substance of the tree, which ever after cured shrew-struck animals by the touch of a leafy branch.

The poor creature that was imprisoned, Ariel-like, in the tree, was, fortunately for itself, not gifted with Ariel's powers of life; and the orifice of the hole being closed by the plug, we may hope that its sufferings were not long, and that it perished immediately for want of air. Still, our fathers were terribly and deliberately cruel; and if the shrew's death was a merciful one, no credit is due to the authors of it.

For on looking through a curious work on natural history, of the date of 1658, where each animal is treated of medicinally, I find recipes of such terrible cruelty that I refrain from giving them, simply out of tenderness for the feelings of my reader. Torture seems to be a necessary medium of healing; and if a man suffers from "the black and melancholy cholic," or "any pain and grief in the winde-pipe or throat," he can only be eased therefrom by medicines prepared from some wretched animal in modes too horrid to narrate, or even to think of.

We are not quite so bad at the present day; but still no one with moderate feelings of compassion can pass through our streets without being greatly shocked at the wanton cruelties practised by human beings on those creatures that were intended for their use, but not as mere machines. Charitably, we may hope that such persons act from thoughtlessness, and not from deliberate cruelty; for it does really seem a new idea to many people that the inferior animals have any feelings at all.

When a horse does not go fast enough to please the driver, he flogs it on the same principle that he would turn on steam to a locomotive engine, thinking about as much of the feelings of one as of the other.

Much of the present heedlessness respecting animals is caused by the popular idea that they have no souls, and that when they die they entirely perish. Whence came that most preposterous idea? Surely not from the only source where we might expect to learn about souls—not from the Bible; for there we distinctly read of "the spirit of the sons of man"; and immediately afterwards of "the spirit of the beast," one aspiring, and the other not so. And the necessary consequence of the spirit is a life after the death of the body. Let any one wait in a frequented thoroughfare for only one short hour, and watch the sufferings of the poor brutes that pass by. Then, unless he denies the Divine Providence, he will see clearly that unless these poor creatures were compensated in another life, there is no such quality as justice.

It is owing to sayings such as these, that men come to deny an all-ruling Providence, and so become infidels. They don't examine the Scriptures for themselves, but take for granted the assertions of those who assume to have done so, and seeing the falsity of the assertion, naturally deduce therefrom the falsity of its source. If a man brings me a cup of putrid water, I naturally conclude that the source is putrid too. And when a man hears horrible and cruel doctrines, which are asserted by theologians to be the religion of the Scriptures, it is no wonder that he turns with disgust from such a religion, and tries to find rest in infidelity. In such a case, where is the fault?

All created things in which there is life, *must* live for ever. There is only one life, and all living things only live as being recipients; so that as that life is immortality, all its recipients are immortal.

If people only knew how much better an animal will work when kindly treated, they would act kindly towards it, even from so low a motive. And it is so easy to lead these animals by kindness, which will often induce an obstinate creature to obey where the whip would only confirm it in its obstinacy. All cruelty is simply diabolical, and can in no way be justified.

Supposing that the two cases could be reversed for just one hour, what a wonderful change there would be in the opinion of men; for it may be assumed that the person most given to inflicting pain and suffering is the least tolerant of it himself.

There is, perhaps, hardly one of my readers who does not know some one person who finds an exquisite delight in hurting the feelings of others by various means, such as ridicule, practical jokes, ill-natured sayings, and so on. If so, he will be tolerably certain to find that the same person is especially thin-skinned himself, and resents the least approach to a joke of which he is the subject.

So, if the shrew were to be the afflicted individual, and the human the victim, there would be found no one so averse to the medicinal process as he who had formerly resorted to it under different circumstances.

This principle is finely carried out, in the terrible scene of Dennis, the executioner's, last hours in *Barnaby Rudge*.

These are not pleasant subjects; and we will pass on to another shrew that is generally found in the water, and called from thence the Water-shrew. It is a creature that may be found in many running streams, if the eyes are sharp enough to observe it, and is well worth examination. As it dives and runs along the bottom of the stream, it appears to be studded with tiny silver beads, or glittering pearls, on account of the air-bubbles that adhere to its fur. I have seen a whole colony of them disporting themselves in a little brooklet not two feet wide, and so had a good opportunity of inspecting them.

WATER-SHREW.

I may mention here, as has been done in one or two other works, that nothing is easier than to watch animals or birds in their state of liberty. All that is required is perfect quiet. If an observer just sits down at the foot of a tree, and does not move, the most timid creatures will come within a few yards as freely as if no human being were within a mile. If he can shroud himself in branches or grass or fern, so much the better; but quiet is the chief essential.

It is impossible to form an idea of the real beauty of animal life, without seeing it displayed in a free and unconstrained state; and more real knowledge of natural history will be gained in a single summer spent in personal examination, than by years of book study.

The characters of creatures come out so strongly; they have such quaint, comical, little ways with them; such assumptions of dignity and sudden lowering of the same; such clever little cheateries; such funny flirtations and coquetries, that I have many a time forgotten myself, and burst into a laugh that scattered my little friends for the next half-hour. It is far better than a play, and one gets the fresh air besides.

These little water-shrews are most active in their sports and their work, for which latter purpose they make regular paths along the banks. And as to their sport, they chase one another in and out of the water, making as great a splash as possible, whisk round roots, dodge behind stones, and act altogether just like a set of boys let loose from the school-room. And then—what a revulsion of feeling to see a stuffed water-shrew in a glass-case!

Now for a few words respecting the distant relation of the shrews, namely, the mole. Of its near relation, the hedgehog, there will not be time to speak.

Every one is familiar with the little heaps of earth thrown up by the mole, and called mole-hills. But as the animal itself lives almost entirely underground, comparatively little is known of it; at all events, to the generality of those who see the hills. The mole is not often seen alive; and few who see it suspended among the branches by the professional killer would form any conception of the real character of this subterranean animal.

Meek and quiet as the mole looks, it is one of the fiercest, if not the very fiercest of animals; it labours, eats, fights, and loves as if animated by one of the furies, or rather by all of them together.

MOLE.

Intervals of profound rest alternate with savage action; and according to the accounts of country folks near Oxford, it works and rests at regular intervals of three hours each.

Useful as these creatures are as subsoil drain-makers, they sometimes increase to an inconvenient extent, and then the professed mole-catcher comes into practice, and destroys the moles with an apparatus apparently inadequate to such a purpose. But the mole is easily killed, and pressure he cannot survive; so the traps are formed for the purpose of squeezing the mole, not of smashing or strangling him.

The mole-catchers are in the habit of suspending their victims on branches, mostly of the willow or similar trees; but their object I could never make out, nor could they give me any reason, except that it was the custom.

When a mole is taken out of the ground, very little earth clings to it. There is always some on its great digging claws; but very little indeed on its fur, which is beautifully formed to prevent such accumulation. The fur of most animals "sets" in some definite direction, according to its position on the body; but that of the mole has no particular set, and is fixed almost perpendicularly on the creature's skin, much like the pile of velvet. Indeed the mole's fur has much the feel of silk-velvet; and so the title of the "Little gentleman in the velvet coat" is justly applied.

Those small heaps of earth that are so common in the fields, and called mole-hills, are merely the result of the mole's travelling in search of the earth-worms, on which it principally feeds; and in their structure there is nothing remarkable.

But the great mole-hill, or mole-palace, in which the animal makes its residence, is a very different affair, and complicated in its structure. In it is found a central chamber in which the mole resides; and round this chamber there run galleries or corridors in a regular series, so as to form a kind of labyrinth, by means of which the creature may make its escape, if threatened with danger.

The accompanying cut shows a section of the mole-palace.

MOLE-HILL.

This palace is formed, if possible, under the protection of large stones, roots of trees, thick bushes, or some such situation; and is located as far as possible from paths or roads.

The food of the mole mostly consists of earth-worms, in search of which it drives these tunnels with such assiduity. The depth of the tunnel is necessarily regulated by the position of the worms; so that in warm pleasant days or evenings the run, as it is called, is within a few inches of the surface; but in winter the worms retire deeply into the unfrozen soil, and thither the mole must follow them. For this purpose it sinks perpendicular shafts, and from thence drives horizontal tunnels. It may be seen how useful this provision is when one thinks of the work that is done by the mole when providing for its own sustenance.

In the cold months, it drives deeply into the ground, thereby draining it, and preventing the roots of plants from becoming sodden by the retention of water above; and the earth is brought from below, where it was useless, and, with all its properties inexhausted by crops, is laid on the surface, there to be frozen, the particles to be forced asunder by the icy particles with which it is filled, and, after the thaw, to be vivified by the oxygen of the atmosphere, and made ready for the reception of seeds.

The worms have a mission of a similar nature; but their tunnels are smaller, and so are their hills. Every floriculturist knows how useful for certain plants are the little heaps of earth left by the worms at the entrance of their holes. And by the united exertions of moles and worms a new surface is made to the earth, even without the intervention of human labour.

Among other pets, I have had a mole—rather a strange pet, one may say; but I rather incline to pets, and have numbered among them creatures that are not generally petted—snakes, to wit—but which are very interesting creatures, notwithstanding.

Being very desirous of watching the mole in its living state, I directed a professional catcher to procure one alive, if possible; and after a while the animal was produced. At first there was some difficulty in finding a proper place in which to keep a creature so fond of digging; but the difficulty was surmounted by procuring a tub, and filling it half full of earth.

In this tub the mole was placed, and instantly sank below the surface of the earth. It was fed by placing large quantities of earth-worms or grubs in the cask; and the number of worms that this single mole devoured was quite surprising.

As far as regards actual inspection, this arrangement was useless; for the mole never would show itself, and when it was wanted for observation, it had to be dug up. But many opportunities for investigating its manners were afforded by taking it from its tub, and letting it run on a hard surface, such as a gravel-walk.

There it used to run with some speed, continually grubbing with its long and powerful snout, trying to discover a spot sufficiently soft for a tunnel. More than once it did succeed in partially burying itself, and had to be dragged out again, at the risk of personal damage. At last it contrived to slip over the side of the gravel-walk, and, finding a patch of soft mould, sank with a rapidity that seemed the effect of magic. Spades were put in requisition; but a mole is more than a match for a spade, and the pet mole was never seen more.

I was by no means pleased at the escape of my prisoner; but there was one person more displeased than myself—namely, the gardener: for he, seeing in the far perspective of the future a mole running wild in the garden, disfiguring his lawn and destroying his seed-beds, was extremely exasperated, and could by no blandishments be pacified.

However, his fears and anxieties were all in vain, as is often the case with such matters, and a mole-heap was never seen in the garden. We therefore concluded that the creature must have burrowed under the garden wall, and so have got away.

Sometimes the fur of the mole takes other tints besides that greyish black that is worn by most moles. There are varieties where the fur is of an orange colour; and I have in my own possession a skin of a light cream colour.

A perpetual thirst seems to be on the mole, for it never chooses a locality at any great distance from water; and should the season turn out too dry, and the necessary supply of water be thus diminished or cut off, the mole counteracts the drought by digging wells, until it comes to a depth at which water is found.

I should like to say something of the Hedgehog, the Stoat, and other wild animals; but I must only take one more example of the British Mammalia, the common Weasel.

WEASEL.

Gifted with a lithe and almost snake-like body, a long and yet powerful neck, and with a set of sharp teeth, this little quadruped attacks and destroys animals which are as superior to itself in size as an elephant to a dog.

Small men are generally the most pugnacious, and the same circumstance is noted of small animals. The weasel, although sufficiently discreet when discretion will serve its purpose, is ever ready to lay down that part of valour, and take up the other.

Many instances are known of attacks on man by weasels, and in every case they proved to be dangerous enemies. They can spring to a great distance, they can climb almost anything, and are as active as—weasels; for there is hardly any other animal so active: their audacity is irrepressible, and their bite is fierce and deep. So, when five or six weasels unite in one attack, it may be imagined that their opponent has no trifling combat before him ere he can claim the victory. In such attacks, they invariably direct their efforts to the throat, whether their antagonist be man or beast.

They feed upon various animals, chiefly those of the smaller sort, and especially affect mice; so that they do much service to the farmer. There is no benefit without its drawbacks; and in this case, the benefits which the weasel confers on farmers by mouse-eating is counterbalanced, in some degree, by a practice on the part of the weasel of varying its mouse diet by an occasional chicken, duckling, or young pheasant. Perhaps to the destruction of the latter creature the farmer would have no great objection.

The weasel is a notable hunter, using eyes and nose in the pursuit of its game, which it tracks through every winding, and which it seldom fails to secure. Should it lose the scent, it quarters the ground like a well-trained dog, and occasionally aids itself by sitting upright.

Very impertinent looks has the weasel when it thus sits up, and it has a way of crossing its fore-paws over its nose that is almost insulting. At least I thought so on one occasion, when I was out with a gun, ready to shoot anything—more shame to me! There was a stir at the bottom of a hedge, some thirty yards distant, and catching a glimpse of some reddish animal glancing among the leaves, I straightway fired at it.

Out ran a weasel, and, instead of trying to hide, went into the very middle of a footpath on which I was walking, sat upright, crossed its paws over its nose, and contemplated me steadily. It was a most humiliating affair.

The weasel has been tamed, and, strange to say, was found to be a delightful little animal in every way but one. The single exception was the evil odour which exudes from the weasel tribe in general, and which advances from merely being unpleasant, as in our English weasels, to the quintessence of stenches as exhibited by the Skunk and the Teledu. A single individual of the latter species has been known to infect a whole village, and even to cause fainting in some persons; and the scent of the former is so powerful, that it almost instantaneously tainted the provisions that were in the vicinity, and they were all thrown away.

The Polecat, Ferret, Marten, and Stoat belong to the true weasels; the Otters and Gluttons claiming a near relationship.

CHAPTER III.

THE COMMON LIZARD—SUDDEN CURTAILMENT—BLIND-WORM—A CURIOUS DANCE—THE VIPER—CURE FOR ITS BITE—THE COMMON SNAKE—SNAKE-HUNTING—CURIOUS PETS—SNAKE AND FROG—CASTING THE SKIN—EGGS OF THE SNAKE—HYBERNATION—THE FROG—THE TADPOLE—THE EDIBLE FROG—THE TOAD—TOADS IN FRANCE—TOAD'S TEETH—VALUE OF TOADS—MODE OF CATCHING PREY—POISON OF THE TOAD—CHANGE OF ITS SKIN.

I have already said that the birds must be entirely passed over in this little work; and therefore we make a jump down two steps at once, and come upon the Reptiles, of whom are many British examples.

The first reptile of which we shall treat is the common little Lizard that is found in profusion on heaths, or, indeed, on most uncultivated grounds.

THE COMMON LIZARD.

It is an agile and very pretty little creature, darting about among the grass and heather, and twisting about with such quickness that its capture is not always easy. Sunny banks and sunny days are its delight; and any one who wishes to see this elegant little reptile need only visit such a locality, and then he will run little risk of disappointment.

There is one peculiarity about it that is rather startling. If suddenly seized, it snaps off its tail, breaking it as if it were a stick of sealing wax, or a glass rod. Several lizards possess this curious faculty, and of one of them we shall presently treat.

The food of this lizard is composed of insects, which it catches with great agility as they settle on the leaves or the ground. If captured without injury— a feat that cannot always be accomplished, on account of the fragility of its tail—it can be kept in a fern case, and has a very pretty effect there.

One of the chief beauties of this animal is its brilliant eye; and this feature will be found equally beautiful in many of the reptiles, and especially in that generally-hated one, the toad.

In the winter-time the lizard is not seen; for it is lying fast asleep in a snug burrow under the roots of any favourable shrub, and does not show itself until the warm beams of the sun call it from its retreat.

The next British lizard that I shall mention is one that is generally considered as a snake, and a poisonous one; both ideas being equally false. It is popularly known by the name of the Blind-worm, or Slow-worm; and is not a snake at all, but a lizard of the Skink tribe, without any legs.

BLIND-WORM.

The scientific name for it is *Anguis fragilis*; and it is called fragile on account of its custom of snapping itself in two, when struck.

Only very lately, I saw an example of this strange propensity, and was the cause of it. Near Dover, there is a small wood, where vipers are reported to dwell; and as I was walking in the wood, I caught a glimpse of a snake-like body close by my foot. I struck, or rather stabbed, it with a little stick—for it had a very viperine look about it—and with success rather remarkable, for the very slight blow that the creature could have received from so insignificant a weapon, used in such a manner. The viper was clearly cut into two parts, but how or where could not be seen, owing to the thick leaves and grass that rose nearly knee-high.

On pushing among the leaves, I found with regret that the creature was only a blind-worm.

A curious performance was being exhibited by the severed tail, a portion of the animal about five inches long; this was springing and jumping about with great liveliness and agility, entirely on its own account, for by this time the blind-worm itself had made its escape, and all search was unavailing.

Some ten minutes or so were consumed in looking for the reptile itself; and by that time the activity of the tail was at an end, and it was lying flat on the ground, coiled into a curve of nearly three-fourths of a circle. I gave it a push with the stick, when I was startled by the severed member jumping fairly into the air, and recommencing its dance with as much vigour as before. This performance lasted for some minutes, and was again exhibited when the tail was roused by another touch from the stick. Nearly half-an-hour elapsed before the touch of the stick failed to make the tail jump, and even then it produced sharp convulsive movements.

The object of this strange compound of insensibility and irritability may perhaps be, that when an assailant's attention is occupied by looking at the tail, the creature itself may quietly make its escape.

The food of the blind-worm is generally of an insect nature, and it seems to be fond of small slugs. The country people declare that it is guilty of various crimes, such as biting cattle and similar offences, of which bite an old author says that, "unless remedy be had, there followeth mortality or death, for the poyson thereof is very strong".

Fortunately for us, we have but one poisonous reptile, the viper; and the slow-worm is as innocent of poison as an earthworm. It is true that, if provoked, it will sometimes bite; but its mouth is so small, and its teeth so minute, that it cannot even draw blood.

The names that are given to it are hardly in accordance with its formation, for it is not very sluggish in its movements, although it can be easier taken than the lizard; while it is anything but blind, and its eyes, though small, are brilliant. Perhaps the epithets ought to have been applied to the givers, and not to the receiver.

As for the real snakes, there are but two species in England, one being called the Viper, or Adder, and the other the Ringed, or Grass-snake. The Viper is rather to be avoided, as it is possessed of poison-fangs, and if irritated, is not slow in using them.

Of this latter I have little to say, and would not have mentioned it excepting for two reasons: the one to enable any person to distinguish it from the common snake, and to avoid, as far as possible, the chance of being bitten; and the other to tell how to heal the bite, should so untoward an event happen.

Poisonous snakes may be readily known by the shape of their head and neck; the head being very wide at the back, and the neck comparatively small. Some persons compare the head of a poisonous snake to the ace of spades, which comparison, although rather exaggerated, gives a good idea of the poison-bearing head. It has a cruel and wicked look about it also, and one recoils almost instinctively.

VIPER.

Should a person be bitten by the viper, the effects of the poison may be much diminished by the liberal use of olive oil; and the effect of the oil is said to be much increased by heat. Strong ammonia, or hartshorn, as it is popularly called, is also useful, as is the case with the stings of bees and wasps, and for the same reason. The evil consequences of the viper's bite vary much in different persons, and at different times, according to the temperament of the individual or his state of health.

I may as well put in one word of favour for the viper before it is dismissed. It is not a malignant creature, nor does it seek after victims; but it is as timid as any creature in existence, slipping away at the sound of a footstep, and only using its fangs if trodden on accidentally, or intentionally assaulted.

The second English snake is the common harmless Ringed Snake; which does not bite, because it has no teeth to speak of; and does not poison people, because it has no venom at all.

COMMON SNAKE.

Its only mode of defence is by pouring forth a most unpleasant, pungent odour, which adheres to the hands or clothes so pertinaciously, that many washings are required before it is expelled. Yet it is sparing enough even of this solitary weapon, and may, after a while, be handled without any inconvenience.

To this assertion I can bear personal and somewhat extensive witness; for I have caught and kept numbers of snakes. The worthy villagers must have formed curious ideas of me, and I rather fancy must have accredited me with something of the wizard character; for I contrived to oppose their prejudices—all, by the way, of a cruel character—in so many instances, that they were rather afraid, as well as annoyed. To see them run away, as if from a lighted shell, when I came among them with a snake in each hand, was decidedly amusing, and not less curious was the pertinacity with which they clung to their prejudices.

In vain were arguments used to prove that the snake was not a venomous animal, and ought not to be killed and tortured; in vain did I put my finger into the snake's mouth, and let its forked tongue glide over my very hand or face; they were not to be so taken in, and they remained wise in their own conceit.

They certainly could not deny that the snake did not bite me, and that its tongue did not pierce me, but the conclusion deduced therefrom was simply that my constitution, or perchance my magical art, was such that I was unbitable and unpoisonable.

No! to them the snake was still poisonous, and its tongue still envenomed.

At one time we had so many snakes that they were kept in the crevices of an old wall, and left to stay or go as they pleased. My boys—I had a school at that time—took wonderfully to snake hunting, and every half-holiday produced a fresh supply of snakes. The boys used to devise the strangest amusements in connection with their snakes, of which they were very proud, each boy exhibiting his particular favourite, and expatiating on its excellences.

One of their fashions, and one which lasted for some time, was to make tunnels in the side of the Wiltshire Downs, and to turn in their snakes at one end, merely for the purpose of seeing them come out at the other.

Then there was a stone-quarry some three miles distant, which was in some parts of the year nearly filled with water. Thither the boys were accustomed to repair for the purpose of indulging their snakes with a bath. They certainly seemed to enjoy the swim, and were the better for it.

Sometimes there was great excitement; for a snake would now and then act in too independent a manner, and instead of swimming straight across, so as to be caught by a boy on the opposite side, would sink to the bottom, and there lie flat and immovable. Long sticks could not be found there; and their only mode of making the snake stir was to startle it by throwing stones. Even then there was a difficulty; for if the stones fell too far from the snake they had no effect, and if they fell on him they might hurt him.

To wait until the truant chose to move would have been hopeless, for snakes are able to take so much pure air into their lungs, and they require so little of it for respiration, that the patience of the boys would be exhausted long before the snake felt a necessity for moving.

Sometimes a snake would try to get away, and insinuate his head and part of his body into a crevice; in that case there was sad anxiety, and judicious management was required to eliminate the reptile without damage. It is a very difficult matter to drag a snake backwards, because the creature sets up the edges of the scales, and each one serves as a point of resistance. So, when the snake is within a crevice, where the scales of the back can act as well as those of the belly, the difficulty is increased.

When such an event took place, the best mode of extracting the snake was to let it glide on, and so lower its scales, and then to pluck it out with a sudden jerk, before it had time to erect them afresh. But as often as otherwise, the snake got the better in the struggle, and by slow degrees was lost to view.

Perhaps the pleasantest portion of snake-keeping was the feeding. It was found that the snakes lost their appetite, and would not eat, though frogs and newts were liberally supplied. So the boys settled the matter by opening the mouth of the snake, and pushing a newt fairly down its throat.

One of the largest snakes that I have seen was engaged in feeding himself, not trusting to boys for any help. I was walking in a field, and heard a strange cry from a neighbouring ditch. On going towards the spot, I saw there a large snake struggling with a frog. The frog was comparatively as large as the snake, and as it had a plain objection to being swallowed, there was some turmoil.

The snake was stretched along the bottom of the ditch, which at this time was dry, and he held in his mouth both hind feet of the frog, who was also stretched forward at full length, resisting with its fore-legs the attempts of the snake to draw it back, and croaking dismally. The strife continued for some time, when I made a sudden movement, and the snake, loosing its hold of the frog, glided up the opposite bank. The frog slowly gathered itself together, sat still for some little time, and then hopped away.

The entire empty skin of the snake may often be found among bushes, where the creature has gone in order to assist itself in casting off its old skin. Snakes, as well as other animals, wear out their coats, and are obliged to change them for others. When the change is about to take place, and a new coat has formed under the old, like a new skin under a blister, the creature betakes itself to some spot where is thick grass, reeds, or similar substances. A rent then opens in the neck, and the snake, by wriggling about among the stems, literally crawls out of its skin, which it leaves behind, turned inside out. Even the covering of the eyes is cast away, and in consequence the snake is partially blind for a day or two previously to the moult, if we may call it so.

Eggs laid by the snake are also of frequent occurrence. I have found them in manure heaps, the warmth of which places is attractive to them. The eggs are white, and covered with a strong membrane, but have no shell. They are laid in long strings, from sixteen to twenty eggs being in each chain.

In the winter the ringed snake retires to a convenient cell, such as a hollow tree, or a heap of wood, and there it remains in a torpid state until the warm weather. Many individuals have been found collected together in these winter quarters, probably for the sake of affording each other mutual warmth.

The reptiles of which we have just treated live exclusively on land, though they may occasionally be found in water; but those which we shall now inspect belong rather to the water than to the land. The most common of these amphibious reptiles, as they are called, is the Frog.

A very curious animal is a frog, and well worth examining, as well in its perfect state as in its intermediate state. To begin at the beginning of a frog's existence, we find it exhibited in masses of eggs, fixed to each other by a kind of gelatinous substance, and floating in large quantities in ditches or ponds. Each egg is about the size and shape of a pea, and in the centre is the little black speck from which the young frog proceeds.

FROG.

In process of time the egg is hatched, and out comes a queer little creature, with a big head and a flat slender tail, called generally a tadpole, and in some places a pollywog. In this state of life the young creature is simply a fish, with fish-like bones, and breathing through gills, after the manner of fish.

Being very voracious, it grows rapidly: little legs begin to show themselves; and, at the proper season, the gills are laid aside, the tail vanishes, and the little frog is then in its usual form. The circulation of the blood can be well exhibited by means of a microscope, if a tadpole be laid on the stage so as to bring its tail within the focus, care being taken to keep that member well wetted.

At the time when the tail is laid aside, the young frog is very small, and in this state is generally found to swarm immediately after rain. The frog-showers, of which we so often hear, are probably occasioned, not by the actual descent of frogs from the clouds, but from the genial influence of the moisture on the young frogs who have already been hatched and developed, and who have been biding their time before they dared to venture abroad.

Still I would not venture to say that frogs have not descended *in* the rain, for there are several accredited accounts of fish-showers, both being probably caused in the same way.

For a drawing of the Tadpole, see page 85.

It is not often that frogs are found far from water, for they are the thirstiest of beings, and drink with every pore of their body. If, for example, a wrinkled and emaciated frog is placed in confinement, and plentifully supplied with water, it absorbs the grateful moisture like a sponge, and plumps up in a wonderfully short time.

From the same cause, it parts with its moisture with equal rapidity; and if a dead frog be laid in the open air on a dry day it speedily shrinks up, and becomes hard as horn. The skin and lungs co-operate in respiration, but only when the former is moist. So, in order to secure that object, the frog is furnished with an internal tank, so to speak, which receives the superabundance of the absorbed water, and keeps it pure until it is required for use. So great is the power of absorption that a frog has been known to absorb a quantity of water equal to itself in weight, merely through the pores of the abdominal surface, and this in a very short time.

In England we don't eat frogs, for what reason I know not. One species of frog is very excellent food, and it is but natural to suppose that another may be so, *i.e.*, if properly cooked. However, the old belief still keeps its ground, that the French are the natural foes of the English, and we ought to hate them, because they "eat frogs and are saddled with wooden shoes". Still I cannot but think that to eat frogs is better than to starve or to steal, and that to wear wooden shoes is not more humiliating than to wear no shoes at all.

After its fashion, the frog sings, though it is but after a fashion. We call the frog's song a croak: I wonder what name the frog would give to our singing. When the frog sings, it generally sinks itself under water, with the exception of its head, opens its mouth, lays its lower jaw flat on the water, and sets to work as if it meant to make the best of its time. Even in England we have fine specimens of frog concerts, though not to such an extent as in many other countries. In France the frogs make such a croaking, that we hardly wonder at the rather tyrannous conduct of the noblesse just before the great Revolution. When the nobility or courtiers spent any time in the country, the miserable peasants were forced to flog the water all night, on purpose to keep the frogs quiet, for their croaking was so noisy that the fastidious senses of the fashionables could not be lulled to sleep.

Now-a-days, the people don't seem to be satisfied with the country croakings, but they import the horrid sounds into the city by means of a toy called a "grenouille," which, when set in motion, makes a croaking sound just like that of a frog.

As a general fact, frogs are just endurable, and people will inspect them—from a distance—without much ado. But the case is widely altered when they see the frog's first-cousin, the Toad.

A large volume might easily be filled with tales respecting this much-calumniated creature; in which tales the toad appears to be a very incarnation of malignity, and to be wholly formed of poison. If it burrowed near the root of a tree, every one who ate a leaf of that tree would die; and, if he only handled it, would be struck with sudden cramp. And the cause of this poisonous nature was its liver, which was "very vitious, and causeth the whole body to be of an ill temperament".

Fortunately, toads had two livers; and although both of them were corrupted, yet one was full of poison, and the other resisted poison. As for remedies, the only effectual one was of rather a complicated nature, and consisted of plantain, black hellebore, powdered crabs, the blood of the sea-tortoise mixed with wine, the stalks of dogs' tongues, the powder of the right horn of a hart, cummin, the vermet of a hare, the quintessence of treacle, and the oil of a scorpion, mixed and taken *ad libitum*.

THE COMMON TOAD.

Even in the days when this prodigious prescription was invented, some good was acknowledged to exist in a toad, the one being the precious jewel in its head, and the other its power as a styptic. Supposing any one to fall down and knock his nose against a stone, he could instantly stop the bleeding if he only had in his pocket a toad that had been pierced through with a piece of wood and dried in the shade or smoke. All that was requisite was to hold the dried toad in the hand, and the bleeding would immediately cease. The reason for this effect is, that "horror and fear constrained the blood to run into his proper place, for fear of a beast so contrary to humane nature".

And, as a concluding instance of the wonderful things that happened whenever toads were the subject, we are told that at Darien, where the household slaves water the door-steps in the evening, all the drops that fall on the right hand turn into toads.

These poor creatures fare little better even now, as far as public opinion goes; and in France worse than in England.

I was once walking in the forest at Meudon with a party of friends, and was brought to a check by a sudden attack made on a large toad that was walking along the pathway. I succeeded in stopping a blow that was aimed at it; and was stooping down, intending to remove it to a place of safety, when I was hastily pulled away, and horror was depicted on the countenances of all the spectators.

"It will bite you," cried one.

"Pouah!" exclaimed another, "it will spit poison at you."

"In France, every one kills toads," said a third.

I objected that it could not bite, because it had no teeth.

"No teeth!" they all exclaimed. "In France, toads *always* have teeth."

"Well, then," I said, "I will open its mouth, and show you that it has none."

But before I could touch it I was again dragged away.

"Teeth come when the toads are fifty years old," was the explanation that was given; but still the death-sentence had passed in every mind, and I knew that when I moved the poor toad would be killed.

Just then, some one remarked that tobacco killed toads, if put on their backs. So I took advantage of the assertion, and made a compromise that, on my part, I would not handle the toad; and that, on theirs, the only mode by which they might kill it was by putting tobacco on it.

The terms being thus arranged, plenty of tobacco was produced—and very bad tobacco, too, as is generally the case in France; and, as no one but myself dared come so near, I put about half-an-ounce of the weed on the back of the toad, as it sat in a rut. For a minute or more, the creature sat quite still, and all the party began to exclaim:—

"See! the toad is quite dead!"

"Ah! the nasty animal!"

"Monsieur Ool!—(no one ever made a better shot at my name than Ool)—Monsieur Ool! the toad is dead!"

However, the toad rose, shook off all the tobacco, and recommenced his march along the road. The only good that was done was the saving of that individual toad's life, for all the party retained their faith in toads' teeth, and probably thought that the creature would not touch me because I was a trifle madder than the rest of my nation, who are always very mad on the French stage.

Afterwards, I found that the belief in toads' teeth was quite general; and one person offered to show me some, half-an-inch in length, which he kept in a box at home. But I was never fortunate enough to see them.

In England, toads are sometimes valued for the good which they do; and the market-gardeners, whose trim grounds surround London, actually import toads from the country, paying for them a certain sum per dozen. For toads are voracious creatures, feeding upon slugs, worms, grubs, and insects of various kinds, and so devour great numbers of these little pests to the gardener.

The mode in which a toad catches its prey is curious enough. Its tongue is fastened into its mouth in a very peculiar way, the base of the tongue being fixed at the entrance of the mouth, the tip pointing down its throat when it is at rest. When, however, the toad sees an insect or slug within reach, the tongue is suddenly shot out of the mouth, and again drawn back, carrying the creature with it.

So rapidly is this operation performed, that the insect seems to disappear by magic. The frog feeds in the same manner.

For the poisonous properties attributed to the toad, there is some foundation, though a small one. But a very small foundation is generally found strong enough to bear a very large superstructure of calumny; though the reverse is the case when the report is a favourable one. The skin of the toad is covered with small tubercles, which secrete an acid humour sufficiently sharp and unpleasant to prevent dogs from carrying a toad in their mouths, though not so powerful as to deter them from attacking toads and killing them.

A rather curious advantage has been taken of the insect-eating propensities of the toad. A gentleman had killed a toad at a very early hour one morning, and after skinning it, for the purpose of stuffing the skin, he dissected its digestive system. The contents of the stomach he turned out into a basin of water, and found there a mass of insects, some of them very rare and in good preservation.

Afterwards, he was accustomed to kill toads for the express purpose of collecting the insects that were found within them, and which, being caught during the night, were often of such species as are not often found.

The same experiment elicited another curious fact, namely, the great tenacity of life possessed by some insects. Before pinning out the insects that were found, and which were mostly beetles, they had been allowed to remain in the water for several days, and were apparently dead. Yet, when they were pinned on cork, they revived; and, when they were visited, were found sprawling about in quite a lively style.

Like all the reptiles, the toad changes its skin, but the cast envelope is never found, although those of the serpents are common enough. The reason why it is not found is this: the toad is an economical animal, and does not choose that so much substance should be wasted. So, after the skin has been entirely thrown off, the toad takes its old coat in its two fore-paws, and dexterously rolls it, and pats it, and twists it, until the coat has been formed into a ball. It is then taken between the paws, pushed into the mouth, and swallowed at a gulp like a big pill.

CHAPTER IV.

NEWTS—A FISH WITH LEGS—NEWTS FEEDING—NEWT-CANNIBALS—CASTING THE SKIN—STRANGE STORIES—ANOTHER NEWT STORY—HATCHING OF YOUNG—TENACITY OF LIFE—THE STICKLEBACK—ITS PUGNACITY—ITS COLOURS—ACCLIMATISATION—THE LAMPERN—A RUSTIC PHILOSOPHER—THE CRAY-FISH—HOW WE CAUGHT IT—REPRODUCTION OF LIMBS—FRESH-WATER SHRIMP—WOODLOUSE AND ARMADILLO.

The Newts, or Efts, or Evats, as they are called in different parts of England, can be easily distinguished from the lizard by the flattened tail, which, being intended for swimming, is formed accordingly.

THE COMMON NEWT.

Two species of these creatures are found in this country, the common Water-Newt and the Smooth Newt. These beautiful creatures may be found in almost every piece of still water, from ponds and ditches up to lakes. The full beauty of the newt is not seen until the breeding season begins to come on, and even then only in the male.

At this time the green back and orange belly attain a brighter tint, and the back is decorated with a wavy crest, tipped with crimson. This crest is continually waving from side to side as the creature moves, and forms graceful curves. The newts are equally at home in water and on land, and in the latter case have often been mistaken for lizards.

One of these animals, when taking a walk, alarmed an acquaintance of mine sadly. He was rather a tall man than otherwise, and did not appear particularly timid; but one day he came to me looking somewhat pale, and announced that he had just been terribly frightened.

"A fish, with legs!" said he, "*four* legs! got out of the water and ran right across the path in front of me! I saw it run!"

"A fish with legs!" I replied; "there are no such creatures."

"Indeed there are, though, for I saw them. It had FOUR LEGS, and it waggled its tail! It was horrible, horrible!"

"It was only a newt," I replied, "an eft. There is nothing to be afraid of."

"It was the *legs*," said he, shuddering, "those dreadful legs. I don't mind getting bitten, or stung, but I can't stand legs."

Newts are very interesting animals, though they have legs, and can easily be kept in a tank if fed properly. Little red worms seem to be their favourite food, and the newt eats them in a rather peculiar style. I have had numbers of newts of all sizes and in all stages of their growth, and always found them eat the worm in the same way. As the worm sank through the water, the newt would swim to it, and by a sudden snap seize it in the middle. For nearly a minute it would remain with the worm in its mouth, one end protruding from each side of its jaws. Another snap would then be given, and after an interval a third, which generally disposed of the worm.

When they have been swimming freely in a large pond, I have often seen large newts attack the smaller, and try to eat them; but I never saw the attempt successful, though I hear that they have been seen to devour the younger individuals. They always came from behind, as if trying to avoid observation, and then made a sudden dart forward, snapping at the tail of their intended victim. In confinement I never saw even an attempt at cannibalism.

Whether it is invariably the case I cannot say, but every newt that I took cast its skin within a few hours from the time that it was placed in the glass jar. The general surface of the skin came off in flakes, but that from the paws was drawn off like gloves, retaining on their surface all the markings and creases which they exhibited when in their proper place.

How the drawing off of their tiny gloves was effected I could not see, though I watched carefully. They looked beautiful as they floated in the water, being delicate as gossamer, white, and almost transparent. They might have been made for Queen Mab herself, and were so delicate that I never could preserve any of them so as to give a proper idea of their form.

It may be that the change of water might cause the change of skin, for the water in which they were kept was drawn from a pump, and that in which they formerly lived was the ordinary soft water found in ponds.

Pretty as is the newt, it is as harmless as pretty, and notwithstanding has suffered under the reputation of being a venomous creature. The absurd tales that I have heard of this creature could scarcely be believed; and how people with any share of sense could receive such absurdities is matter of wonder. And as usual, the moral of the stories is, that newts are to be killed wherever found. The belief of the poisonous character of the newt is of long standing, as may be seen in the ancient works on natural history. In one of these it is said that its poison is like that of vipers; and there is a description of the formation of its tail which is rather beyond my comprehension:—

"The tail standeth out betwixt the hinder-legs in the middle, like the figure of a wheel-whisk, or rather so contracted as if many of them were conjoined together, and the void or empty places in the conjunctions were filled".

The capture and domesticating of newts gave dire offence in the village where I lived for some time; and the expressions used when I took a newt in my hands were not unlike those of the Parisians respecting the toad. Sundry ill-omened tales of effets were told me. For example: A girl of the village was filling her pitcher at a stream which runs near the village, when an effet jumped out of the water, sprang on her arm, bit out a piece of flesh, spat fire into the wound, and, leaping into the water, escaped. The girl's arm instantly swelled to the shoulder, and the doctor was obliged to cut it off.

This was told me with an immensity of circumstantial details common to such narrators, and was corroborated by the bystanders. The wounded lady herself was not to be found, and cross-questions elicited that it "weir afoor their time". I asked them how the effet which lived in the water, and had just leaped out of it, was able to keep a fire alight in its interior; but they were not in the least shaken, except perhaps in their heads, which were wagged with a Lord Burleigh kind of emphasis.

Then there was the sexton-clerk-gardener-musician and general factotum, who had a newt tale of his own to tell. He had been cutting grass in the churchyard, and an effet ran at him, and bit him on the thumb. He chopped off the effet's head with his knife, but his thumb was very bad for a week.

Once they got the better of the argument, at all events in the eyes of the owner of the farming stock, and my poor newts were ejected. It happened thus:—

Two or three specimens I kept in my own room in a glass vase, in order to watch them more closely; and some six or seven others lived as stock in the large horse-trough, from whence they could be taken when required.

One day the proprietor came to me and ordered the destruction of my newts, for they had killed one of his calves.

"But," I remonstrated, "they cannot kill a calf or even a mouse, for they have no fangs and very little mouths. Besides, the calf has not come near this trough."

So saying, I took up several of the newts, opened their mouths—no easy matter, by the way—and showed that they had no fangs. And I urged, that even if they had been as poisonous as rattlesnakes, it would not have made any difference to the calf, which had never left the cowhouse, and was at the opposite end of the farm-yard, separated by a barn and several gates. But all was useless.

"There are the newts, and there is the dead calf!" was the answer; and so the newts had to go. However, I would not suffer them to be killed, but put them into a bag and took them back to the pond whence they had come.

Afterwards the proprietor said that the calf died because its mother had drunk at the trough in which the poisonous newts were.

Now, the funniest part of the story is, that there was not a horse-pond that did not swarm with efts, and consequently all the foals and calves ought to have died. Only they didn't.

The care which the female newt takes in depositing her eggs is very remarkable.

THE FEMALE NEWT.

Each egg is taken separately, and by the aid of the fore-paws is regularly tied or twisted up in the leaves of dead plants, for which process different people have different reasons. Some think that it is for the purpose of preventing too ready an access of water, and so to retard their hatching; while some say that it is to guard the egg against voracious water-animals. To the latter opinion I rather incline; perhaps both may be right.

When hatched, the young newt is very like a tadpole, breathing by gills outside its neck. After a while the gills vanish and the legs appear; but it keeps its tail. It is rather curious that the frog tadpole puts forth its hinder-legs first; while in the tadpole of the newt, the fore-legs are the first to show themselves.

After the gills are lost, the newt breathes by means of lungs; and if it is in the water, is forced to rise at intervals for the purpose of breathing.

The tenacity with which these creatures cling to life is quite surprising. Experiments have been tried purposely to see to what degree a body could be mutilated, and yet retain life. They have even been frozen up into a solid block of ice, and, after the thawing of their cold prison, revived, and seemed none the worse for it. I may as well mention that none of these experiments were tried by myself, for I am not scientific enough to care nothing for the infliction of pain; but on one occasion I did try an experiment, and, as it turned out, a very cruel one, although it was not intended for an experiment.

I was studying the anatomy of the frogs and newts; and having eight or ten fine specimens of the latter creature, determined to take advantage of the opportunity. The first thing was, of course, to kill the creature without injuring its structure, and I thought that the best mode of so doing would be to put it into my poison-bottle. This was a large glass jar filled with spirits of wine, in which was held corrosive sublimate in solution. This mixture generally killed the larger insects almost immediately, and seemed just the thing for the newts.

So they were put into the jar—but then there was a scene which I will not describe, which I trust never to see again, and of which I do not even like to think. Suffice it to say, that nearly a quarter of an hour elapsed before these miserable creatures died, though in sheer mercy I kept them pressed below the surface.

Changing our post of observation from the banks of the ponds to those of the running streams, we shall find there many creatures worthy of observation; so many, indeed, that it would be a hopeless task to attempt to give even a slight account of one-fiftieth of them. I shall, therefore, only mention two creatures, as examples of the fish; and these two are chosen because they are exceedingly common, and very different from each other in colour and habits.

The first of these creatures is the common Stickleback, or Tittlebat, as it is sometimes called. There are several species of British sticklebacks; but the commonest, and I think the most beautiful, is the three-spined stickleback.

These little fish derive their name from the sharp spines with which they are armed, and which they can raise or depress at pleasure—as I know to my cost. For being, as boys often are, rather silly, I made a wager that I would swallow a minnow alive; and having made the bet, proceeded to win it. Unfortunately, instead of a minnow, a stickleback was handed to me, which having its spines pressed close to the body, was very like a minnow. Just as I swallowed it, the creature stuck up all its spines, and fixed itself firmly.

THE STICKLEBACK.

Neither way would it go, and the torture was horrid. At last, a great piece of apple that I swallowed gave it an impetus that started it from its position; but it was not for some time, that to me appeared hours, that the fish was disposed of. And even then it left its traces; and if it would be any satisfaction to the fish to know that ample vengeance was taken for its death, it must have been thoroughly gratified.

There are few fish more favoured in point of decoration than the stickleback; although the decoration, like that of soldiers, is only given to the gentlemen, and of them only to the victors in fight.

They are most irritable and pugnacious creatures, that is, in the early spring months, when the great business of the nursery is in progress. And the word nursery is used advisedly; for the stickleback does not leave her eggs to the mercy of the waters, but establishes a domicile, over which her husband keeps guard.

The vigilance of this little sentry is wonderful; and I have often seen fierce fights taking place. Not a fish passes within a certain distance of the forbidden spot, but out darts the stickleback like an arrow, all his spines at their full stretch, and his body glowing with green and scarlet. So furious is the fish at this time that I have sometimes amused myself by making him fight a walking-stick.

If the stick were placed in the water at the distance of a yard or so, no notice was taken. But as the stick was drawn through the water, the watchful sentinel issued from his place of concealment, and when the intruding stick came within the charmed circle, the stickleback shot at it with such violence that he quite jarred the stick.

His nose must have suffered terribly. If the stick were moved, another attack would take place, and this would be continued as long as I liked.

Sometimes a rival male comes by, with all *his* swords drawn ready for battle, and his colours of red and green flying. Then there is a fight that would require the pen of Homer to describe. These valiant warriors dart at each other; they bite, they manœuvre, they strike with their spines, and sometimes a well-aimed cut will rip up the body of the adversary, and send him to the bottom, dead.

When one of the combatants prefers ignominious flight to a glorious death, he is pursued by the victor with relentless fury, and may think himself fortunate if he escapes.

Then comes a curious result. The conqueror assumes brighter colours and a more insolent demeanour; his green is tinged with gold, his scarlet is of a triple dye, and he charges more furiously than ever at intruders, or those whom he is pleased to consider as such. But the vanquished warrior is disgraced; he retires humbly to some obscure retreat; he loses his red, and green, and gold uniform, and becomes a plain civilian in drab.

Sometimes I have brought on a battle royal between the guardians of several palaces, by dropping in the midst of them a temptation which they could not resist. This was generally a fine fat grub taken from a caddis case. The caddis is large and white, and so can be seen to a considerable distance.

As this sank in the water, there would be a general rush at it, and the ensuing contention was amusing in the extreme. First, one would catch it in his mouth and shoot off; half-a-dozen others would unite in chase, overtake the too fortunate one, seize the grub from all sides, and tug desperately, their tails flying, their fins at work, and the whole mass revolving like a wheel, the centre of which was the caddis worm.

It would be swallowed almost immediately, but the mouth of the stickleback is much too small to admit an entire caddis, and the skin of the grub is too tough to be easily pierced or torn. Half-an-hour often elapses before the great question is settled, and the caddis eaten.

The rapidity of the evolutions and the fierceness of the struggle must be seen to be appreciated—and it is a spectacle easily to be witnessed; wherever there are sticklebacks, caddis worms are nearly certainly found, and it only needs to extract one of these from its case and deposit it judiciously in the water.

The stickleback is a hardy little fish, and can easily be kept in the aquarium, if plenty of room be given to it. It has even been trained to live in seawater, by adding bay-salt to the water in which it dwelt; so that the plan of pickling salmon alive, by a judicious admixture of vinegar and allspice with the water, has something to which to appeal as collateral evidence.

The other representative of the fishes is a very curious one, and can be easily observed. It is called the "Lampern," and is shown in the accompanying figure.

THE LAMPERN.

In some parts of England the lampern goes by the name of "Seven-eyes," in allusion to the row of eye-like holes that may be seen extending along the side of the throat. These apertures are the openings by which the water passes from the gills.

The chief external peculiarity in this creature is the mouth, which, instead of being formed with jaws like those of other fishes, resembles none of them, not even those of the eel, which it most resembles externally. Indeed, on looking at the mouth of a lampern, one is forcibly reminded of the leech, for it is possessed of no jaws, and adheres firmly to the skin by exhaustion of the air.

Very delicate food are these lamperns, quite as good as the lampreys themselves, whose excellence is reported to have cost England one of her kings; yet I never knew but one person who would eat them, and very few who would even touch them, they also being called poisonous.

In Germany they know better, and not only eat the lamperns themselves, but, packing them up in company with vinegar, bay leaves, and spices, export them as an article of sale.

The solitary sensible individual of whom I have made mention was truly a wise man. He used to offer the young urchins of the neighbourhood a reward for bringing lamperns, at the rate of a halfpenny per wisketful.

A wisket, I may observe, is a kind of shallow basket, made of very broad strips of willow; and a wisket filled with lamperns would be a tolerable load for a boy.

So, for the sum of one halfpenny, that philosopher was furnished with provisions for a day or more.

Really, the prejudice against the lampern is most singular. Even near London, when lamperns lived in vast numbers in the Thames, they were only used as bait, being sold for that purpose to the Dutch fishermen. In one season, four hundred thousand of these creatures have been sold merely for bait for cod-fish and turbot.

The scientific name for the lampreys is "*Petromyzon*," a word signifying "stone-sucker". The name is rightly applied; for when the lampern wishes to remain still in one place, it applies its mouth to a stone, sticks tightly to it by suction, and there remains firmly at anchor, and defying the power of the stream. In favourable spots, thousands of these fish may be seen together, quite blackening the bottom of the stream with their numbers. They seem specially to affect shallow mountain streams; and, in spite of the rapid current, wriggle their devious way up the stream with great rapidity. When they are not quite pleased with the spot on which they settle down for the time, they scoop it out to their minds in a very short time. This task is accomplished by means of the sucker-like mouth. If a stone is placed in a position that incommodes them, they affix their mouths to it, and drag it away down the stream. In this way they will remove stones which are apparently beyond the power of so small a creature. By perseverance they thus scoop out small hollows, about eighteen inches long and a foot wide, in which they lie in groups so thick that I have more than once mistaken them for dark logs lying in the stream, and was only undeceived by the waving of the multitudinous tails. Year after year the lamperns followed the same course, and chose the same positions, so that we could at any time tell where these creatures would be found by the thousand, where they would be found singly, and where none would be seen at all.

The general thickness of this creature is that of a large pencil, but it varies according to the individual. The length is from one foot to fifteen inches or so.

There is a much smaller species of lampern called the Pride, Sand-pride, or Mud Lamprey, which is not more than half the length of the lampern, and only about the thickness of an ordinary quill. This creature has not the power of affixing itself like the lampern, on account of the construction of its mouth.

Having now taken a hasty glance at the vertebrated animals, we pass to those who have no bones at all, and whose skeleton, so to speak, is carried outside. Our representation of aquatic crustacea, as such creatures are called, will be the Cray-fish and the Water-Shrimp.

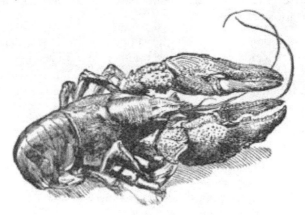

THE CRAY-FISH.

Every one knows the Cray-fish, because it is so like a lobster, turning red when boiled in the same way. This red colour is brought out by heat even if applied by placing the shell before a fire, and spirits of wine has the same effect. The last fact I learned from experience, and was very sorry that it *was* a fact, for the red shell quite spoiled the appearance of a dissected cray-fish that was wanted to look nice in a museum.

Being very delicate food, and, in my opinion, much better than the native lobster, they are much sought after at the proper season, and are sold generally at the rate of half-a-crown for one hundred and twenty.

There are many modes of catching them, which may be practised indifferently. There are the "wheels," for example, being wicker baskets made on the wire mouse-trap principle, which the cray-fish enters and cannot get out again. Also, there is a mode of fishing for them with circular nets baited with a piece of meat. A number of these nets are laid at intervals along the river bank, and after a while are suddenly pulled out of the water, bringing with them the cray-fish that were devouring the meat.

But the most interesting and exciting mode of cray-fish catching is by getting into the water, and pulling them out of their holes.

Cray-fish take to themselves certain nooks and crannies, formed by the roots of willows or other trees that grow on the bank; and they not unfrequently take possession of holes which have been scooped by the water-rat. The hand is thrust into every crevice that can be detected, and if there is a cray-fish, its presence is made known by the sharp thorny points of the head,—for the cray-fish always lies in the hole with its head towards the entrance.

The business is, then, to draw the creature out of its stronghold without being bitten—a matter of no small difficulty. If the hole is small, and the cray-fish large, I always used to draw it forward by the antennæ or horns, and then seize it across the back, so that its claws were useless.

The power of the claws is extraordinary, considering the size of the creature that bears them. They will often pinch so hard as to bring blood; and when they have once secured a firm hold, they do not easily become loosened. Still, the risk of a bite constitutes one of the chief charms of the chase.

The legitimate mode of disposing of the cray-fish, when taken, is to put them into the hat, and the hat on the head; but they stick their claws into the head so continually, and pull the hair so hard, that only people of tough skin can endure them.

Sometimes, when the bed of the river is stony, the cray-fish live among and under the stones, and then they are difficult of capture; for with one flap of their tail they can shoot through the water to a great distance, and quite out of reach.

It is not unfrequent to find a cray-fish with one large claw and the other very small. The same circumstance may be noted in lobsters. The reason of this peculiarity is, that the claw has been injured, generally in single combat; for the cray-fish are terrible fighters, and the mutilated limb has been cast off. Most wonderfully is this managed.

The blood-vessels of the crustaceans are necessarily so formed, that if wounded, they cannot easily heal; and if there were no provision against accidents, the creature might soon bleed to death.

But when a limb, say one of the claws, is wounded the limb is thrown off—not at the injured spot, but at the joint immediately above. The space exposed at the joints is very small in comparison with that of an entire claw; and as the amputation takes place at a spot where there is a soft membrane, it speedily closes. In process of time, a new limb begins to sprout, and takes the place of the member that had been thrown off.

The eyes of the cray-fish are set on footstalks, so as to be turned in any direction, and they can also be partially drawn back, if threatened by danger. If the eye is examined through a magnifying glass of tolerable power, it will be seen that it is not a single eye, but a compound organ, containing a great number of separate eyes, arranged in a wonderful order. As, however, a description of an insect's eye will be given at a succeeding page, we at present pass over this organ.

At the proper season of the year, the female cray-fish may be seen laden with a large mass of eggs, which she carries about with her, and by the movement of the false legs that are arranged in double rows on the under surface of the tail, keeps them supplied with fresh streams of water. In process of time, the eggs are hatched; but very few, in comparison, reach maturity. Even the mother herself is apt to eat her own young, when they have set themselves free from her control. I have known this to take place when we were trying to breed cray-fish in a tank. Only one attained to any size, and even that was not so large as a house-fly when we took it from the water.

FRESH-WATER SHRIMP.

TADPOLES AND YOUNG FROG.

The fresh-water Shrimp may generally be found in plenty in any running stream. Its appearance and habits very much resemble the Sandhopper, a little creature that every one must have seen who has walked on a sandy sea-shore. Like the cray-fish, this little creature carries its eggs about until they are hatched. It is a carnivorous animal, and is one of the numerous scavengers of the water, without whose help every stream would soon become putrid and loathsome.

WOODLOUSE, ARMADILLO, AND PILL MILLEPEDE.

Certain species of crustacea inhabit the land; two of which are well known under the titles of Woodlouse and Armadillo. They belong to the class of crustaceans called "*Isopod,*" or equal-footed, because the legs are all of the same nature; whereas, in the other crustacean, some legs are used for walking, and others are turned into claws, &c. The woodlouse is to be found in myriads under the scaly bark of trees, under stones, and, in fact, in almost every crevice. It feeds mostly on decayed vegetable matters, but also eats animal substances, and vegetables that are not decayed. Some gardeners hold the woodlouse in great horror, and say that nothing is so hard or so bitter that a woodlouse will not eat it. If the bark is removed from an ancient willow tree, any number of these creatures may be discovered, in every stage of existence, scuttling about in great fear at the unwelcome light, and sticking close to the wood in hopes that they may not be seen. Dried coats of the woodlouse may be also seen, empty and bleached to an ivory whiteness. They are night-feeders; and, although they can run fast enough if disturbed, walk very deliberately when only employed in feeding.

The Armadillo-woodlouse is very curious, and easily recognised from its habit of rolling itself into a round ball when alarmed, just like the quadruped armadillo. Its habits are much the same as those of the common woodlouse. Formerly the armadillo was used in medicine, being swallowed as a pill in its rolled-up state. I have seen a drawer half full of these creatures, all dry and rolled up, ready to be swallowed.

On the preceding cut are two armadillo-like animals, much resembling each other, but belonging to different orders. Fig. *a* is the Woodlouse; *b*, the Pill Millepede, walking; *c*, the same rolled up; *d* is the true Armadillo, walking; and *e*, the same creature rolled up.

One of the minute crustaceans, the common Cyclops, is shown on plate J, fig. 14. Its length is about the fourteenth of an inch; and though it is so small, the female Cyclops may easily be detected in the water by the curious egg-sacs.

CHAPTER V.

A SHORT ESSAY ON LEGS—TAKING A WALK—BRITISH FAKIRS—INSECT LIFE—DEVELOPMENT—THE TIGER MOTH—GROWTH OF THE CATERPILLAR—HOW TO DISSECT INSECTS—PLAN OF CATERPILLAR ANATOMY—SILK ORGANS—ORGANS OF RESPIRATION—SPIRACLES AND THEIR USE—WONDERS OF NATURE—THE CHRYSALIS—SCIENTIFIC LANGUAGE.

As, in common with many other animals, mankind are furnished with legs, and the power to move them, it is universally acknowledged that those limbs ought to be put to their proper use. But while men agree respecting the importance of the members alluded to, they differ greatly in the mode of employing them.

To the tailor, for example, legs are chiefly valuable as cushions, whereon to lay his cloth. For the jockey, the same members form a bifurcated or pronged apparatus, by the help of which he sticks on a horse. The legs of the acrobat are mostly employed to show the extent of ill-treatment to which the hip-joint can be subjected without suffering permanent dislocation. The dancer values his leg solely on account of the "light fantastic toe" which it carries at its extremity. The turner sees that two legs are absolutely necessary to mankind—*i.e.*, one to stand upon, and the other to make a wheel run round. The surgeon views legs—on other people—as objects affording facilities for amputation. The boxer professionally regards his legs as "pins," upon which the striking apparatus is kept off the ground. The soldier's opinion of his legs is modified according to the temperament of the individual, and the position of the enemy. Some people employ their legs in continually mounting the same stairs, and never getting any higher; while others use those limbs in continually pacing the same path and never going any farther.

And of all these modes of employing the legs, the last, which is called "taking a walk," is the dreariest and least excusable.

For, in the preceding cases, the owners of the legs gain their living, or at all events their life, by such employment of those members; and in the case of the interminable stairs, the individual is not acting by his own free will. But it does seem wonderful that a being possessed of intellectual powers should fancy himself to be the possessor of a right leg and a left one, merely that the right should mechanically pass the left so many thousand times daily and in its turn be passed by the left; while the sentient being above was occupied in exactly the same manner as if both legs were at rest, snugly tucked under a table.

Sad to relate, such is the general method of taking recreation.

A man who has been over-tasking his brain all the early part of the day, rises corporeally from his work at a certain time, places his hat above his brain, buttons his coat underneath it, and sallies forth to take a walk.

Whatever subject he may be working upon he takes with him, and on that subject he concentrates his attention. Supposing him to be a mathematician, and that the prevalent idea in his mind is to prove that $\triangle \, A\,B\,C = (\angle \, D\,E\,F + \angle \, G\,H\,I)$. He takes one final look at his Euclid while drawing on his gloves, and sets off with A B C before his eyes.

As he walks along, he sees nothing but A B C, hears nothing but D E F, feels nothing but G H I, and thinks of nothing but the connection of all three.

An hour has passed away and he re-enters his room without any very definite recollection of the manner in which he got there. He has mechanically paced to a certain point, mechanically stopped and turned round, mechanically retraced his steps, and mechanically come back again.

He has not the least recollection of anything that happened during his walk; he don't know whether the sky was blue or cloudy, whether there was any wind, nor would he venture to say decidedly whether it was night or day. He *does* recollect seeing a tree on a hill and a spire in a valley, because, together with himself, they formed an angle that illustrated the proportions of the triangle A B C; but whether the tree had leaves or not he could not tell. But he is happy in the consciousness of having performed his duty;—he has taken a walk, he has been for a "constitutional".

O deluded and misguided individual! The walking powers are meant to carry yourself—not only your corporeal body—into other scenes, to give a fresh current to your thoughts, and to give your brain an airing as well as your nose. The mind requires variety in its food, as does the body; and to obtain that change of nutriment is the proper object of taking a walk.

That a rational being can condemn himself to walk three miles along a turnpike road, and three miles back again, at one uniform pace, his eyes directed straight ahead, and his thoughts at home with his books, seems incredible to ordinary personages.

Yet, such British fakirs may be seen daily in all weathers, on the roads leading from university towns, going at the rate of four miles per hour, their hats tilted towards the back of their heads, their bodies inclining forward at an angle of eighty degrees, their lips muttering polysyllabic language, and their eyes as beaming as those of a boiled cod-fish.

Now the real use of taking a walk is to get away from one's self, and to change the current of the thoughts for a while, by changing the locality of the individual.

In order so to do, he should cast his senses abroad instead of concentrating them all within himself; and from sky, air, water, and earth draw a new succession of images wherewith to relieve the monotony within. There are various modes of attaining this object; and each man will follow that mode which most accords with his own character.

For example, if he is an astronomer, he will look to the heavenly bodies; if a geologist, his eyes will be directed to the earth; if a botanist, his mind seeks employment among the vegetable productions; if a meteorologist, the wind's temperature and atmospheric phenomena will claim his attention; if an entomologist, he will find recreation in watching the phases of insect life, and so on.

It is evident enough that to treat of all these subjects would render necessary a volume that numbered its pages by thousands, and its volumes by at least tens; and therefore, in a work of this nature, it must be sufficient to lay particular stress on one portion, to treat slightly of others, and to leave many entirely untouched. And that portion on which I shall lay the chief stress is that which is brought more constantly before the eye and ear than any other, namely, the entomological department.

As, when approaching cities, the "busy hum of men" is the first indication that meets the ear, so in the country the busy hum of insects is, next to the song of the birds, the sound that gives strongest evidence of a life untrammelled by the artificial rules of society.

Not only do insects make their presence known to the ear, but they also address themselves to the eye. Their forms may be seen flitting through the air, running upon the ground, or making their abode on the various examples of vegetable life. Comparatively small as insects are, they are of vast importance collectively; and there is hardly a leaf of a tree, a blade of grass, or a square inch of ground, where we may not trace the work of some insect. Nearly all strange and curious objects that are noticed by observant eyes in the woods or fields are caused by the action of insects, and are often the insects themselves, in one or other of the phases of their varied life. Certain examples of insect life, and its effects, will now be given. No particular order will be observed, no long scientific terms will be used, and every creature that is mentioned will be so common that it may be found almost in every field.

The first creature that we will notice is that caterpillar which is so abundantly found at several seasons of the spring and summer, and, from the long hairy skin in which it is enveloped, goes by the popular name of the "Woolly Bear!" A figure of this creature may be seen in plate B, fig. 5 *a*. This creature is the larva of the common Tiger-moth, which is represented on the same plate, fig. 5.

It will be necessary to pause here a little, before proceeding to the description and histories of the various insects, because in the course of description certain terms must be used, which must be explained in order to make the description intelligible.

In the first place, let it be laid down as a definite rule, that

INSECTS NEVER GROW.

Many people fancy that a little fly is only little because it is young, and that it will grow up in process of time to be as big as a blue-bottle. Now this idea is entirely wrong; for when an insect has once attained to its winged state, it grows no more. All the growing, and most part of the eating, is done in its previous states of life; and, indeed, there are many insects, such as the silkworm-moth, which do not eat at all from the time that they assume the chrysalis state to the time when they die.

It is a universal rule in nature, that nothing comes to its perfection at once, but has to pass through a series of changes, which if carefully examined can mostly be reduced to three in number. Sometimes these changes glide imperceptibly into each other, but mostly each stage of progress is marked clearly and distinctly. Such is the case with the insect of which we are now considering; and when we have examined the development of the Tiger-moth through its phases of existence, we have the key to the remainder of the insects.

After an insect has left the egg, and entered upon the world as an individual being, it has to pass through three stages, which are called larva, pupa, and imago.

The word "larva," in Latin, signifies "a mask," and this word is used because the insect is at that time "masked," so to speak, under a covering quite different from that which it will finally assume. In the present instance, the Tiger-moth is so effectually masked under the Woolly Bear, that no one who was ignorant of the fact would imagine two creatures so dissimilar to have any connection with each other.

Throughout this work the word "larva" will be always understood to signify the first of the three states of insect life, whether it be a "caterpillar," a "grub," or a "worm".

In its next stage the insect becomes a "pupa," which word means a "mummy," or a body wrapped in swaddling clothes. This name is employed because in very many insects the pupa is quite still, is shut up without the power of escape, and looks altogether much like a mummy, wrapped round in folds of cloth. In the moths and butterflies the insect in this stage is called a "chrysalis," or "aurelia," both words having the same import, the first Greek and the other Latin, both derived from a word meaning "gold". Several butterflies—that of the common cabbage butterfly, for example—take a beautiful golden tinge on their pupal garments, and from these individual instances the golden title has been universally bestowed.

The last, and perfected state, is called the "imago," or image, because now each individual is an image and representative of the entire species.

The Woolly Bear, then, is the larva of the Tiger-moth; and if any inquiring reader would like to keep the creature, and watch it through its stages, he will find it an interesting occupation. There is less difficulty than with most insects, for the creature is very hardy, and the plant on which it mostly feeds is exceedingly common.

Generally, the Woolly Bear is found feeding on the common blind nettle, but it may often be detected at some distance from its food, getting over the ground at a great rate, and reminding the spectator of the porcupine. In this case it is usually seeking for a retired spot, whither it resorts for the purpose of passing the helpless period of pupa-hood.

If it is captured on such an occasion, there will be little trouble in feeding, as it will generally refuse food altogether, and, betaking itself to a quiet corner, prepare for its next stage of existence.

If taken at an earlier period of its life, it feeds greedily on the nettle above-mentioned, and the amount of nutriment which one caterpillar will consume is perfectly astounding. I once had nearly four hundred of them all alive at the same time, and they used to be furnished with nettles by the armful. Of course so large a number is not necessary for ordinary purposes; but this regiment was required for the purpose of watching the development and anatomy of the creature through its entire life.

As the skins of caterpillars are not capable of growth, and the creature itself grows with singular rapidity, it is evident that the skins themselves must be changed, as is the case with many other animals of a higher class, such as the snakes, newts, &c.

For this purpose the skin of the caterpillar splits along the back of the neck, and by degrees the creature emerges, soft, moist, and helpless. A very short time suffices for the hardening of the new envelope; and as the caterpillar has been obliged to fast for a day or two, previously to changing the skin, it sets to work to make up for lost time, and does make up effectually.

In the case of the Woolly Bear, and several others, the cast skin retains nearly the same shape and appearance as when it formed the living envelope of the caterpillar; and, consequently, if any number of these insects are kept, the interior of their habitation soon becomes peopled with these imitation caterpillars. Each individual changes its skin some ten or eleven times, each time leaving behind it a model of its former self, so that caterpillars seem to multiply almost miraculously.

Although even the exterior appearance of an insect is very wonderful, yet its interior anatomy is, if possible, even more wonderful, and, if possible, should be examined. The mode of doing so is simple and easy. If the Woolly Bear, for example, is to be dissected, the easiest mode of doing so is as follows:—

Get a shallow vessel, glass if possible, about an inch or so in depth; load a flat piece of cork with lead, put it at the bottom of the vessel, and fill it nearly to the top with water. Now take the caterpillar, which may be killed by a momentary immersion in boiling water, or by being placed in spirits of wine, and with a few minikin pins fasten it on its back on the cork. The pins of course must only just run through the skin, and two will be sufficient at first, one at each end.

Now take a pair of fine scissors, and carefully slit up the skin the entire length of the creature, draw the skin aside right and left, and pin it down to the cork.

The creature will now exhibit portions of organs of different shapes and characters, the remainder being concealed under the mass of fat that is collected in the interior. This fat must be carefully removed in order to show the vital organs; and this object is best attained by using a fine needle stuck into a handle. I generally use a common crochet-needle handle, so that needles of various sizes can be used at pleasure.

Now will appear a number of organs closely packed together, and mostly stretching along the entire length of the creature. In order to assist the inquirer, I here present a plan or chart of the interior of the caterpillar when thus opened. It must be understood that the drawing is not meant to represent the particular anatomy of any one species, but to give a general view, by means of which the anatomical details of any caterpillar may be recognised. And in order to give greater distinctness, only one of each organ is seen, though with the exception of the intestinal canal, there is a double set of each organ, one on each side.

Running in a straight line from head to tail is seen the digestive apparatus, consisting of throat, stomach, and intestines, with their modifications; and this apparatus is marked A A in the cut.

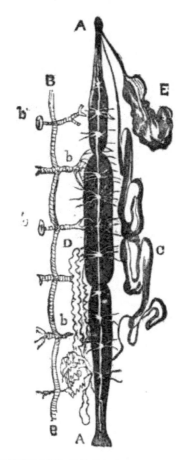

INTERIOR OF CATERPILLAR.

On the surface of the digestive apparatus, and straight along its centre, lies the nervous system, represented by tiny white threads dotted at regular distances by rather larger spots of the same substance. If the nerve is examined closely, it will be seen to be composed of *two* very slender threads, lying closely against each other, but easily separable: in which state they are shown. And the little knobs are called "ganglia," each forming a nervous centre, from which smaller nerves radiate to the different portions of the body.

As for brains, the caterpillar dispenses with them almost entirely; and instead of wearing one large brain in the head, is furnished with a row of lesser brains, or ganglia, extending through its whole length. This is the reason why caterpillars are so tenacious of life. If a man loses his head, he dies immediately; but an insect is not nearly so fastidious, and continues to live for a long time without any head at all. Indeed, there are some insects, which, if beheaded, die, not so much on account of the head, but of the stomach: for, having then no mouth, they cannot eat, and so die of hunger. And some insects there are which positively live longer if decapitated than if left in possession of their head.

On the right hand may be seen a curiously twisted organ, marked C, swelling to a considerable size in the middle, and diminishing to a mere thread at each end. This is one of the vessels that contain the silk, or rather the substance which becomes silk when it is spun.

If this organ be cut open in the middle, it will be seen filled with a gummy substance of curious texture, partly brittle and partly tough. From this substance silk is spun, by passing up the tube, through the thread-like portion, and so at last into a tiny tube, called the spinneret, which opens from the mouth, and wherefrom it issues in a fine thread.

There are two of these silk-making organs, and both unite in the spinneret. Consequently, if silk is examined in the microscope, the double thread can clearly be made out, both threads adhering to each other, but still distinguishable. If the threads lie parallel to each other, the silk is good; if not so, it is of an inferior quality, and liable to snap.

Most caterpillars possess this silk-factory, but some have it much more largely developed than others—the silk-worm, for instance. It is of considerable size in the larva which we are examining, because the Woolly Bear has to spin for itself a silken hammock in which to swing while it is in the sleep of its pupal state. Just before it begins to spin, the organ is of very large size, and distended with the liquid silk; but after the hammock is completed, the organ diminishes to a mere thread, and is soon altogether absorbed.

At the left hand of the drawing may be seen a curious structure, marked B B. This is the chief portion of the respiratory system, and may be at once recognised by the ringed structure of the tube. Indeed it is quite analogous to that of the windpipe in animals.

The mode in which insects breathe differs much from that of the higher animals. In them the breathing apparatus is gathered into one mass, called lungs or gills, as the case may be; but with insects, the respiratory system runs entirely over, round, and through the body, even to the tips of the claws, and the end of the feelers or antennæ.

Every internal organ is also surrounded and enveloped by the breathing tubes; and this often to such an extent, that the dissector is sadly perplexed how to remove the tracheal tubes, as they are called, without injuring the organs to which they so tightly cling. Sometimes they are so strongly bound together, that they may be removed like a net, but mostly each must be taken away separately. The mode in which these tracheal tubes supply the digestive apparatus may be seen at $b\ b$; and as there is a double set of them, it may be seen how closely they envelop the organ to which they direct their course.

The ringed structure runs throughout the entire course of the air tubes, and is caused by a thread running spirally between the two membranes of which the tube is composed. The object of this curious thread is to keep the tube always distended, and ready for the passage of air. Otherwise, whenever the insect bends its flexible body, it would cut off the supply of air in every tube which partook of the flexure of the body.

The structure is precisely similar to that of a spiral wire bell-spring; and so strong is the thread, that I have succeeded in unwinding nearly two inches of it from the trachea of a humble bee.

The air obtains entrance into these tubes, not through the mouth or nostrils, but through a set of oval apertures arranged along the sides of the insect, which apertures are called "spiracles"; and two of them are indicated at $b^*\ b^*$.

In order to prevent dust, water, or anything but air, from entering, the spiracles are defended by an elaborate *chevaux de frise* of hair, or rather quill, so disposed as to keep out every particle that could injure. So powerful are these defences, that, even under the air-pump, I was unable to force a single particle of mercury through them, though a stick will be entirely permeated by the metal, so that if cut it starts from every pore. I kept the creature in a vacuum for three days, then plunged it under mercury, and let in the air. Even then no effect was produced, except that the whole of the stomach and intestinal canal were charged with mercury.

But, though the spiracles are such excellent defences against obnoxious substances, they are not capable of throwing off any substance that may choke them. Consequently, nothing is easier than to kill an insect humanely, if one only knows how; and few things more difficult, if one does not know.

For example, if ladies catch a wasp they proceed to immolate it by snipping it in two with their scissors; a dreadfully cruel process, for the poor creature has still some four or five brains left intact, and lives for many hours. But if a feather is dipped in oil and swept across the body of the creature, it collapses, turns on its back, and dies straightway. For the oil has stopped up the spiracles, and so the supply of air is cut off from every portion of the body at once. The same rule holds good with all insects.

There is yet one more organ to which I must draw attention, and that is the curious bag-shaped object marked E.

Just as the silk is contained in the vessel C, so the saliva is contained in E, and is developed according to the character and habits of the insect. Some insects require a large supply of that liquid, which is used for various purposes, and others require comparatively little. The caterpillar in which these receptacles may be found best developed is the larva of the Goat-moth, which may be easily found within the substance of decaying trees. Of the Goat-moth we may speak in a future page.

If the reader will again refer to the engraving on p. 100, he will see that between the tracheal tube and the digestive apparatus is a curiously waved line, forming two loops in its upper portion, and running into a confused entanglement below. This entanglement, however, is only apparent, for in nature there is no entangling; all is perfect in order.

This wavy line represents one of the numerous thread-like vessels that surround this portion of the digestive apparatus, and are called the biliary vessels, being, in fact, the insect's liver. There is a large mass of these biliary vessels, and they are found so closely entwined among each other, and so encircled with the air tubes, that to separate them is no easy matter. Their microscopic structure is curious, and will repay a careful examination.

In examining the creature for the first time, the dissector will be tolerably sure to damage the organs and unfit it for preservation, and therefore it is best to take such a course for granted, and to make the best of it.

Removing all these vital organs, he should then examine the wonderful and most complicated muscular structure, by which the caterpillar is enabled to lengthen, shorten, twist, and bend its body in almost any direction, and that with such power that many caterpillars are enabled to stretch themselves horizontally into the air, and there to keep themselves motionless for hours together.

Few people have any idea of the wonders that they will find inside even so lowly a creature as a caterpillar—wonders, too, that only increase in number and beauty the more closely they are examined. When the outer form has been carefully made out, there yet remains the microscopical view, and after that the chemical, in either of which lie hidden innumerable treasures.

A very forcible and unsophisticated opinion was once expressed to me, after I had dissected and explained the anatomy of a silk-worm to an elderly friend. He remained silent for some time, and then uttered disconnected exclamations of astonishment.

I asked him what had so much astonished him.

"Why," said he, "it's that caterpillar. It is a new world to me. I always thought that caterpillars were nothing but skin and squash."

Having now seen something of the exterior and interior of the caterpillar, we will watch it as it prepares for its next state of existence.

Hitherto it has been tolerably active, and if alarmed while feeding, it curls itself round like a hedgehog and falls to the ground, hoping to lie concealed among the foliage, and guarded from the effects of the fall by its hairy armour, which stands out on all sides, and secures it from harm. But a time approaches wherein it will have no defence and no means of escape, so it must find a means of lying quiet and concealed. This object it achieves in the following manner.

It leaves its food, and sets off on its travels to find a retired spot where it may sling its hammock and sleep in peace. Having found a convenient spot, it sets busily to work, and in a very short time spins for itself a kind of silken net, much like a sailor's hammock in shape, and used in the same manner. It is not a very solid piece of work, for the creature can be seen through the meshes; but it is more than sufficiently strong to bear the weight of the inclosed insect, and to guard it from small foes.

On plate B, and fig. 5 *b*, the silken hammock is represented, the form of the pupa inside being visible. It casts off its skin for the last time, and instead of being a hirsute and active caterpillar, becomes a smooth and quiescent chrysalis. In this state it abides for a time that varies according to the time of year and the degree of temperature, and at last bursts its earthly holdings, coming to the light of the sun a perfect insect.

When first the creature becomes a chrysalis, its colour is white, and its surface is bathed in an oily kind of liquid, which soon hardens in the air, and darkens in the light.

On one occasion, I watched a Woolly Bear changing its skin, and, seizing it immediately that the task was accomplished, put it into spirits of wine, intending to keep it for observation.

Next day, the spirit was found to have dissolved away the oily coating, and all the limbs and wings of the future moth were standing boldly out.

Before closing this chapter, I must just remark that the absence of scientific terms throughout the work will be intentional, from a wish to make the subject intelligible, instead of imposing. It would have been easy enough to speak of the Woolly Bear as the larva of Arctia Caja; to describe it as a chilognathiform larva, with a subcylindrical body, and no thoracic shield: passing through an obtected metamorphosis, and becoming a pomeridian lepidopterous imago; and to have proceeded in the same style throughout. But as nearly every one who has taken a country walk has seen Woolly Bears, and hardly any one knows what is meant by "chilognathiform," the subject is treated of for the benefit of the many, even at the risk of incurring the contempt of the few.

CHAPTER VI

THE PUSS-MOTH—CURIOUS CATERPILLAR—A STRONG FORTRESS—THE BURNET-MOTH—OAK EGGER—HOW TO KILL INSECTS—TWOFOLD LIFE—VICTIMS OF LOVE—ACUTE SENSES—THE STORY OF INSECT LIFE—DRINKER MOTH—CATERPILLAR BOX—EMPEROR MOTH—TYPE OF THE MOUSE-TRAP.

Just at the right hand of the Tiger-moth, on plate B, may be seen a caterpillar of a very strange and eccentric form, and marked by the number 4 *a*. This is the larva or caterpillar of the Puss-moth, and is no less beautiful in colouring than fantastic in form. Its attitude, too, when it is at rest, is quite as curious as its general appearance.

While eating, it sits on the leaves and twigs much as any other caterpillar; but when it ceases to feed, and reposes itself, it grasps the twig firmly with the claspers with which the hinder portion of its body is furnished, and raises the fore-part of its body half upright. In this attitude it much resembles that of the Egyptian Sphinx, and from this circumstance the moth itself is called a Sphinx. An old gardener was once quite put out of temper by seeing several of these caterpillars for the first time, because they had so consequential an air.

The colouring of this creature varies according to the time of year; but it may be easily recognised by its form alone, which is very peculiar.

One of the most remarkable points in the creature is the forked apparatus at the end of the tail, and which frightens people who do not know the habits of the caterpillar. These forks are black externally, and rather stiff, but are only sheaths for two curious rose-coloured tentacles, which are usually kept hidden, but which may be seen by touching the caterpillar with the point of a needle. When the creature is thus irritated, it will protrude these tentacles from their sheath, and will then strike the part that had been touched.

It is supposed that this apparatus is intended as a kind of whip, wherewith to drive away the ichneumon flies, and other parasites, that inflict such annoyance on many caterpillars.

When this caterpillar proceeds to its pupal state, it makes itself a wonderful fortress—not suspended like that of the Tiger-moth, nor hidden in a dark spot; but it boldly fixes its residence on the exterior of the tree on which it feeds, trusting to its similitude to the bark for concealment, and to the strength of its habitation for safety, even if discovered.

It is furnished with a gummy substance, something after the manner of the silk of the Tiger-moth; but instead of spinning that substance into threads, it uses it in the following manner.

Biting little chips of wood from the bark of the tree, the caterpillar glues them together with this natural cement; and so builds an arched house for itself, much about the size and shape of half a walnut-shell. So strongly compacted is this residence, that rain and wind have no effect on it, and a penknife does not find an easy entrance.

One or two of these caterpillars which I brought home modified their dwellings in a curious manner. One of them nibbled to pieces a portion of a cardboard box, and so made a kind of papier-maché house; while others, who were placed under a glass tumbler, and upon a stone surface, simply made their house of the hardened gum. In this state, it appeared as if it had been made of thin horn, and was so transparent that the chrysalis could be seen through the walls.

The caterpillar is common enough, and may be found on the willow or poplar. And a sharp eye will soon learn to detect the winter house, which to an unpractised eye looks as if it were merely a natural excrescence on the bark.

If one of these habitations is found, the best mode of removing it is to avoid touching the dwelling itself, but to cut away the bark round it; and then, by inserting the point of a stout knife, gently raise up the house, together with the bark on which it is placed. This is one of the modes by which an entomologist may find employment even during the winter months, and others will be mentioned in the course of this work.

The moth itself may be seen figured on plate B, fig. 4. It is called the Puss-moth, on account of the soft furry down with which its body is covered, and it is fancifully thought to resemble the fur of the cat.

It is rather a difficult moth to preserve effectually, as it is apt to become "greasy"—that is, to have its whole beauty destroyed by an oiliness that exudes from the body, and gradually creeps even over the wings. The best preservative is to remove the contents of the abdomen, and stuff it with cotton-wool that has been scented with spirits of turpentine. But even that plan is rather precarious, and the delicate, downy plumage is apt to be sadly damaged during the process of stuffing.

Still keeping to the same plate, and referring to the right-hand corner at the top, a moth of strange aspect will be seen; and immediately below it an object that somewhat resembles the hammock of the Tiger-moth, affixed in a perpendicular instead of an horizontal direction. This moth is called the Burnet-moth, and the hammock is the pupa case of the same insect.

The colouring of this moth is very rich and beautiful. The two upper wings are green, and of a tint so deep that, like green velvet, they almost appear to be black. On each of these wings are several red spots, varying in number according to the species; some wearing six spots, and others only five. The two under wings are of a carmine red, edged with a border of black, in which is a tinge of steely blue. The body is velvety black, with the same blue tint.

The moth is rather local; but when one is found in a field, hundreds will certainly be near.

At the best of times it is not an active insect, and on a cold or a dull day hundreds of them may be seen clinging to the upright grass stems, from which they can be removed at pleasure.

The caterpillar of this beautiful moth keeps close to the ground, and feeds on grasses, the speedwell, dandelion, and other plants. When it is about to become a pupa, it ascends some slender upright plant, generally a grass stem, and then spins for itself the residence which is represented on the plate.

In this state it may be gathered, and placed under a glass shade; and in the summer months the perfect insect will make its appearance. There are some places which it specially favours, and where it may be found in great profusion. At Hastings, for example, the fields about the cliffs were so populated by these moths, that hardly a grass stem was without its Burnet-moth's habitation.

Feeding on the same plant as the Tiger-moth caterpillar, may often be found another caterpillar of a very different aspect. It is very much larger, and instead of presenting an array of stiff bristles, is covered with thick soft hair of a yellowish-brown colour, diversified with stripes of a deep velvety black, arranged so as to resemble the slashed vestments that were so fashionable some centuries ago.

This caterpillar is the larva of the Oak Egger-moth, and is not so remarkable as a caterpillar as for the house which it builds for its pupal residence.

After changing its skin the requisite number of times, the caterpillar ceases to feed, and, proceeding to some convenient spot (generally a faded thorn-branch), spins its temporary habitation. This cocoon, as it is called, is about an inch in length; and into that narrow space the creature contrives to push, not only itself, but also its last and largest skin.

The substance of the cocoon is hard and rather brittle when dry; and in texture somewhat resembles thin brown cardboard. In its substance, and on its surface, are woven many of the hairs with which the caterpillar is furnished. If the cocoon is carefully opened, the chrysalis will be found within, its head towards the spot where the moth is to emerge, and the cast caterpillar-skin crumpled down by its tail.

In course of time, the chrysalis passes through its development, and the egger-moth itself pushes its way out of the cocoon, with wings and body wet and wrinkled, but soon to assume their proper form and strength. The cocoon is shown at plate I, fig. 5 *a*.

Sometimes the cocoon remains unbroken beyond the proper season; and if it is examined, one or two little holes will generally be found in it. These are signs that the egger has met with an untimely fate, and that it has fallen a victim to those scourges of the insect world, the ichneumon flies. Of these creatures we shall speak in a future page, and therefore omit to describe them here. The moth is shown at fig. 5.

If the moth is intended to be killed, and then placed in a cabinet, the use of sulphur must be avoided. It kills the moth, certainly; but it kills the colours also, and quite ruins its appearance. Sulphur is always a dangerous instrument in insect-killing, and should on no account be used. There are many ways of destroying insects humanely, and extinguishing their life as if by a lightning flash; but these modes vary according to the size, sex, and nature of the insect. Some of them I will here mention.

If the insect is a beetle, it may be plunged into boiling water, or into spirits of wine, in which a very little corrosive sublimate has been dissolved. Both modes will destroy the life rapidly, but the former is the better of the two. When walking in the fields or woods, a wide-mouthed, strong bottle, about half full of spirits of wine, is a useful auxiliary, as all kinds of beetles, and even flies and bees, can be put into it; and if dried in a thorough draught, will look as well as before. If this precaution be not taken, all the insects that have long hair, as the humble-bee and others, will lose their good looks, and their hair will be matted together in unseemly elf-locks.

Butterflies, and most of the Diptera, or two-winged flies, can be instantaneously killed by a sharp pinch on the under-surface of the thorax among the legs, as the great mass of nerves is there collected. Many people seem to fancy that the head is the vital part in an insect; and having pinched or run a pin through its head, they think that they have effectually slain the creature, and marvel much to see it lively some twenty-four hours afterwards.

Especially is this the case with the large-bodied moths, whose vitality is quite astonishing. You may even stamp upon them, and yet not crush the life out of that frail casket. If you drive the life out of one-half of the creature, it only seems to take refuge in the other; and then retain a more powerful hold, like a garrison driven into a small redoubt.

It is not at all uncommon to find one of these moths dead and dry as to its wings and limbs, which snap like withered sticks if touched, and yet with so much life in it as to writhe its abdomen if irritated, and to deposit its eggs just as if it were in full activity.

Indeed, so strong is this power that the creature seems to be gifted with a double life, one for itself and the other for its progeny. The former is comparatively weak, and but loosely clings to its home; but the latter intrenches itself in every organ, penetrates every fibre, and, until its great work is completed, refuses to be expelled. So, unless the entire mechanism of the insect be killed, the poor creature may live for days in pain.

Fortunately, there is a mode of so doing; and this is the way of doing it:—

Make a strong solution of oxalic acid, or get a little bottle of prussic acid—it is the better of the two, if you have discretion as beseems a naturalist. Also make a bone or iron instrument, something like a pen, but without a nib. Dip this instrument into the poison as you would a pen, and then you have a weapon as deadly as the cobra's tooth, and infinitely more rapid in its work. Now hold your moth delicately as entomologists hold moths, near the root of the wings. Keep the creature from fluttering; plunge the instrument smartly into the thorax, between the insertion of the first and second pair of legs; withdraw it as smartly, and the effect will be instantaneous. The moth will stretch out all its legs to their full extent; there will be a slight quiver of the extremities; they will be gently folded over each other; and you lay your dead moth on the table.

The reason of this rapid decease is of a twofold nature.

In the first place, the chief nerve mass is cut asunder, and even thus a large portion of the life is destroyed. But the chief breathing tubes are also severed, and a drop of poison deposited at their severed portions. Consequently, at the next inspiration, either the poison itself or its subtle atmosphere rushes to every part, and to every joint of the insect, thus carrying death through its whole substance.

The male insect is very different in appearance to the female, and in general is hardly more than two-thirds of her size. The colours, too, are very different; for in the male insect the wings are partially of a dark chestnut brown, with a light band running round them, as may be seen in the engraving; while in the female the wings are almost entirely of a uniform yellowish brown.

The antennæ, too, of the male are deeply cleft, like the teeth of a comb; while those of the female are narrow, and comparatively slightly toothed.

As is the case with several other moths, the male oak eggers are sad victims to the tender passion, and fall in love not only at first sight, but long before they see the object of their affection at all.

If a female egger is caught immediately after her entrance into the regions of air, and placed in a perforated box near an open window, her unseen charms will be so powerfully felt by gentlemen of her own race that they will flock to the casket that contains their desired treasure, and fearlessly run about it, fluttering their wings, and striving to gain admission. So entirely do they abandon themselves to the captivity of love, that they do not fear the risk of a bodily captivity, and will suffer themselves to be taken by hand, without even an endeavour to escape.

Carry the imprisoned moth into the fields, and even there the eager suitors will arrive from all quarters, and boldly alight on the box while in the hand of the entomologist.

More wonderful must be the influence that can emanate from so small a creature, and extend to so great a distance—an influence which, although entirely inappreciable by any human sense, exercises so potent a sway on all sides, and to so great a distance.

The conditions, too, of this mysterious influence are singularly delicate; for after the moth has once found her mate, she may be placed amid a crowd of gentlemen, and not one will take the least notice of her.

Like the young beauty of the ball-room, who whilom attracted to herself crowds of beaux, that fluttered around her, and contended with each other for a look or a smile of their temporary divinity, but who finds herself deserted by the fickle crowd when her election is made; so our Lady Lasiocampa Quercus, after setting all hearts ablaze for a time, makes happy one favoured individual, is deserted by the many rejected, and left in quiet to the duties of a wife and a mother.

Her married life is but short, for her husband rarely survives his happiness more than a few hours, and she, after making due preparation for the welfare of her numerous family, whom she is never to see, feels that she has fulfilled her destiny, and gives up a life which has now no further object.

There is really something very human in the life even of an insect. Many a life story have I watched in the insect world, which, if transferred to the human world, would be full of interest. There is also one great advantage in the insect life, namely, that as it only consists of a year or two, the events of several successive generations come under the observation of a single historian.

First, a number of tiny, purposeless beings come into the world, spreading about much at random, and seeming to have no other object except to eat. It is but just to them to say that they don't cry, and are always contented with the food that is given them.

They rapidly increase in size, pass through a regular series of childish complaints, which we mass together under this single term, "moulting," but which are probably to their senses as distinct as measles, and chicken-pox, and hooping-cough.

They outgrow a great many suits of clothes in a wonderfully short period; they retire for a time to finish their education; and then come before the world in all the glory of their new attire.

Up to this time they are nearly exactly alike in habits and manners; but, when freed from the trammels that held them, they diverge, each in his appointed way, each exulting for a short space in the buoyancy of youth, and fluttering indeterminately in the new world, but soon settling down to the business for which they were made.

So even in insects a human soul can find a companionship, and a solitary man need never feel entirely alone as long as he can watch the life of a humble moth, and see in that despised creature some manifestations of the same feelings which actuate himself.

And it even seems that, through this companionship, the higher nature communicates itself in some degree to the lower, as is shown by the many instances of men who have tamed spiders and other creatures quite as far removed from the human nature. In such a case it seems very clear that either the higher nature gives to the lower an intelligence not its own, or that it develops powers which would have lain dormant had they not been called forth by the contact of a superior being.

This subject is a very wide one, and well worth following up. But as it runs through the whole creation, and this book is only to consist of a few pages, it must suffice merely to put forth the idea.

To pass to another insect.

On plate E, and fig. 1 and 1 *a*, may be seen an insect which somewhat resembles the oak egger-moth, and is often mistaken for it by inexperienced eyes. This is the "Drinker" moth, remarkable for the thick furry coat which it wears, as a caterpillar and as a moth, and which it employs in the construction of its cocoon. This moth is one of my particular friends; and I have had hundreds of them from the egg to their perfect state. I had quite a large establishment for the education and development of lepidoptera, and especially favoured the tiger-moth, the oak egger, and the drinker.

The caterpillar of this moth is entirely covered with dense hair, even down to the very feet; and by means of this protection it is enabled to brave the winter frost, needing not to pass the cold months in a torpid state. It is a pretty caterpillar, and very easily recognised by the figure. Its chief peculiarities are the two tufts of hair that it bears at its opposite extremities, and the double line of black spots along its sides.

Generally, it feeds on various grasses, but it is not dainty, as are many caterpillars; and I have always found it to eat freely of the same food as the oak egger larva. This caterpillar is seen at fig. 1 *b*.

When alarmed, it loosens its hold of the plant on which it is feeding, rolls itself into a ring, and drops to the ground, hoping to evade notice among the foliage. This habit used to be rather perplexing to me, not because the creature could escape by so well-known a trick, but because it would not go into the box prepared for its reception.

It is necessary to have a box of a peculiar form for the collection of caterpillars. If the lid is raised every time that a fresh capture is made, difficulties increase in proportion to the number of caterpillars. For, when some thirty larvæ are in the box, they all begin to crawl out when the lid is opened; and Hercules had hardly a more bewildering task among the hydra's heads than the entomologist among his captives.

No sooner is the light admitted, than a dozen heads are over the side; and as fast as one is replaced, six or seven more make their appearance. The only remedy is to sweep them all back with a rapid movement of the hand, to shake them all to the bottom, and then to replace the lid as fast as possible. Even with all precaution, caterpillars are crushed; and, besides, they are delicate in their constitutions, and require gentle handling.

So the best plan is to have a tin box made with a short tube, through which the caterpillars can be introduced, and which can be stopped by a cork when the creatures are fairly inside.

Now, although this is a capital contrivance for caterpillars that hold themselves straight, it fails entirely when they curl themselves into a ring and refuse to be straightened. It is as impossible to straighten a rolled-up hedgehog as a caterpillar in a similar attitude; and if force is used in either case, the creature will be mortally injured. However, gentle means succeed when violence fails, with insects as with men. A Bheel robber will steal the bedding from under a sleeping man without waking him; and, by an analogous process, the refractory caterpillar is lodged in his prison before he is fairly awake to his condition.

The entomologist feels a justifiable pride in executing similar achievements; for there is quite as much force of intellect needed to outwit a caterpillar as a quadruped.

When the drinker caterpillar passes into its pupal state, it makes for itself a very curious cocoon, not unlike a weaver's shuttle in shape, being large in the middle, and tapering to a point at each end. The texture is soft and flexible, as if the cocoon were made of very thin felt, and the larval hairs are quite distinguishable on its surface. The moth leaves the cocoon about August. For the cocoon see fig. 1 *c*.

COCOON OF THE EMPEROR MOTH.

I found that few caterpillars are so liable to the attack of ichneumon flies as those of the drinker moth. A cocoon now before me is pierced with thirteen holes from which ichneumon flies have issued, having eaten up the caterpillar. The eggs are shown in fig. 1 *e*.

If the reader will now refer to plate C, the central figure will be found to represent a strikingly handsome moth, called, from its gorgeous plumage, the "Emperor Moth".

Its body is covered with a thick downy raiment, and the wings are clothed with plumage of a peculiarly soft character, which is well represented in the figure. The antennæ, too, are elaborately feathered.

Although the beauty of this insect would entitle it to notice in its perfect state, and the peculiar shape of its larva—(see plate C, fig. 4 *a*)—would draw attention, yet its chief title to admiration lies in the cocoon which it constructs for its pupal existence.

Externally, there is nothing remarkable in the cocoon; and, as may be seen in the same plate, fig. 4 *b*, it is a very ordinary, rough, flask-shaped piece of workmanship. But if the outer covering be carefully removed, or if the cocoon be divided lengthways, a very wonderful structure is exhibited.

The inventor of lobster-pots is not known, and history has failed to record the name of the man who first made wire mouse-traps with conical entrances, into which the mice can squeeze themselves, but exit from which is impossible.

But, though the principle had not been applied to lobsters or mice, it was in existence ages upon ages ago. Before human emperors had been invented, and very probably long before mankind had been placed on our earth, the caterpillar of the emperor moth wove its wondrous cell, and thereby became a silent teacher to the cunning race of mankind how to make mouse-traps and lobster-pots.

For inside the rough outer case, which is composed of silken threads, woven almost at random, and very delicate, is a lesser case, corresponding in shape with its covering, but made of stiff threads laid nearly parallel to each other, their points converging at the small end of the case. See the cut on p. 125.

It will now be seen that the moth when it leaves its chrysalid case can easily walk out of the cocoon, but that no other creature could enter. So within its trapped case the chrysalis lies secure, until time and warmth bring it to its perfection. It breaks from its pupal shell, walks forward, the threads separate to permit its egress, and then converge again so closely that to all appearance the cocoon is precisely the same as when the moth was within.

Now, any observant member of the human race, who had been meditating upon traps, and happened in a contemplative mood to open one of these cocoons, would feel a new light break in upon him, and, Archimedes-like, he would exclaim "Eureka," or its equivalent, "I have found my trap!" Reverse the process, make the converging threads to lead into instead of out of the trap, and the thing is done. "I will make it of wire, put it on my shelf, and I catch mice and rats. I will make it of osier, sink it to the bottom of the sea, and I catch lobsters and crabs. I will lay it in a rapid, and I catch roach and dace; I will place it under the river banks, and then I have cray-fish."

So might he soliloquise on the future achievements of his newly-discovered principle. But unless he had the prophetic afflatus strong within him, never would he imagine that in future times his discovery would catch a monarch and an Elector to boot.

CHAPTER VII.

ELEPHANT HAWK-MOTH—PRIVET HAWK-MOTH—DIGGING FOR LARVÆ—BUFF-TIPP MOTH—GOLD-TAILED MOTH—CASE FOR ITS EGGS—CURIOUS PROPERTY OF ITS CATERPILLAR-VAPOURER MOTH—LEAF-ROLLERS—GREEN-OAK MOTH—ITS CONSTANT ENEMY—LEAF-MINERS—LACKEY MOTH—EGG BRACELETS.

It will be noticed that the insects mentioned in the preceding chapter are mostly remarkable for the cocoons which they construct, and that the peculiarities of the larva and the perfect insect are but casually mentioned. Those, however, which will be noticed in this chapter are chosen because there is "something rare and strange" in the habits and manners of the creatures themselves.

As it will be more convenient to keep to the same plate as much as possible, we still refer to plate G. On casting the eye over the objects there depicted, the strangest and most fantastic shape is evidently that creature which is marked 5 *a*.

The aspect of the creature is almost appalling, and it seems to glare at us with two malignant eyes, threatening the poisoned blow which the horrid tail seems well able to deliver.

Yet this is as harmless a creature as lives, and it can injure nothing except the leaves of the plant on which it feeds. The eye-like spots are not eyes at all, but simply markings on the surface of the skin, and the formidable horn at the tail cannot scratch the most delicate skin.

The creature is in fact simply the caterpillar of a very beautiful moth, represented in fig. 5, and called the Elephant Hawk-moth—elephant, on account of its long proboscis, and hawk on account of its sharp hawk-like wings and flight. The caterpillar may be found in many places, and especially on the banks of streams, feeding on various plants, such as the willow-herbs.

Another kind of hawk-moth is much more common than the elephant, and is represented on plate A; the moth itself at fig. 5, and its caterpillar at fig. 5 *a*.

This is called the Privet Hawk-moth, because the caterpillar feeds on the leaves of that shrub. The colours of both moth and caterpillar are very beautiful, and not unlike in character.

The bright leafy green tint of the caterpillar, and the seven rose-coloured stripes on each side, make it a very conspicuous insect, and raise wishes that tints so beautiful could be preserved. But as yet it cannot be done, for even in the most successful specimens the colours fade sadly in a day or two, and after a while there is a determination towards a blackish brown tint that cannot be checked.

Any one, however, who wishes to try the experiment may easily do so, for there are few privet hedges without their inhabitants, who may keep out of sight, but can be brought tumbling to the ground by some sharp taps administered to the stems of the bushes.

In the winter the chrysalis may be obtained by digging under privet bushes. There the caterpillar resorts, and works a kind of cell in the ground for its reception. It is better not to choose a frosty day for the disinterment, or the sudden cold may kill the insect, and the seeker's labour be lost.

Should it be desirable to capture the larva and to keep it alive the object can be easily attained; for the creature is hardy enough, and privet bushes grow everywhere. In default of privet leaves, it will eat those of the syringa and the ash. When it reaches its full growth, it should be provided with a vessel containing earth some inches in depth. Into this earth it will burrow, and remain there until the moth issues forth.

Care should be taken to keep the earth rather moist, as otherwise the chrysalis skin becomes so hard that the moth cannot break out of its prison, and perishes miserably.

On the same plate, fig. 4, may be seen a moth of a curious shape, very feathery about the thorax, the head being all but concealed by the dense down, and as difficult to find as the head of a Skye-terrier, were not its position marked by the antennæ. This is the Buff-tip Moth, so called on account of the upper wing-tips being marked with buff-coloured scales.

The caterpillar, which is represented immediately above, and marked 4 *a*, is a very singular creature, its habits being indicated by the marks on its skin. As soon as the young caterpillars are hatched, they arrange themselves in regular order, much after the fashion of the dark stripes, and so march over leaf and branch, devastating their course with the same ease and regularity as an invading army in an enemy's land.

When they increase to a tolerably large size, they disband their forces, and each individual proceeds on its own course of destruction. Were it not for the colours which they assume, these creatures would do great damage; but the ground being yellow and the stripes black, the caterpillars are so conspicuous that sharp-sighted birds soon find them out, and having discovered a colony, hold revelry thereon, and exterminate the band.

Comparatively few escape their foes and attain maturity. When they have reached their full age, they let themselves drop from the branches, and when they come to soft ground, bury themselves therein to await their last change. Individuals may often be seen crossing gravel paths, which they are unable to penetrate, and getting over the ground with such speed and in so evident a hurry that they seem to be aware that birds are on the watch and ichneumons awaiting their opportunity.

There is a very pretty moth covered with a downy white plumage even to the very toes, and carrying at the extremity of its tail a tuft of golden silky hair. From this coloured tuft, the creature bears the name of Gold-tailed Moth. It may often be found sticking tightly to the bark of tree stems, its glossy white wings folded roof-like over its back, and the golden tuft just showing itself from the white wings.

This golden tuft is only found fully developed in the female moth, and comes into use when she deposits her eggs. The moth is shown on plate E, fig. 4.

As the eggs are laid in the summer time, they need no guard from cold; but they do require to be sheltered from too high a degree of temperature, and for this purpose the silken tuft is used.

At the very end of the tail the moth carries a pair of pincers, which she can twist about in all directions; and this tool is used for the proper settlement of the eggs. The moth, after fixing on a proper spot, pinches off a tiny tuft of down, spreads it smoothly, lays an egg upon it, covers it over, and finally combs the hair so as to lie evenly. And when she has laid the full complement, she gives the whole mass some finishing touches, like a mother tucking-in her little baby in the bed-clothes, and smoothing them neatly over it.

The egg masses are common enough, and are readily discovered by means of their bright yellow covering.

The caterpillar of this moth is a very brilliant scarlet and black creature, commonly known by the name of the "palmer-worm," and to be found plentifully of all sizes.

People possessed of delicate skins must beware of touching the palmer-worm, or they may suffer for their temerity. I was a victim to the creature for some time before I discovered the reason of my sufferings. And the case was as follows.

Being much struck with the vivid colours of the caterpillar, I was anxious to preserve some specimens, if possible, in a manner that would retain the scarlet and black tints. One mode that seemed feasible was to make a very small snuff-box, as ladies call a rectangular rent, in the creature's skin, to remove the entire vital organs, to fill the space with dry sand, and then, when the skin was quite dry, to pour out all the sand, leaving the empty skin.

After treating six or seven caterpillars in this fashion, I perceived a violent irritation about my face, lips, and eyes, which only became worse when rubbed. In an hour or so my face was swollen into a very horrid and withal a very absurd mass of hard knobs, as if a number of young kidney potatoes had been inserted under the skin.

Of course, I was invisible for some days, and after returning to my work, was attacked in precisely the same manner again. This second mischance set me thinking; and on consultation with the medical department, the fault was attributed to the hot sand which I had been using.

So, when I went again to the work, I discarded sand, and stuffed the caterpillars with cotton wool cut very short, like chopped straw. My horror may be conjectured, but not imagined, when I found, for the third time, that my face was beginning to assume its tubercular aspect.

Then I did what I ought to have done before, went to my entomological books, and found that various caterpillars possessed this "urticating" property, as they learnedly called it, or as I should say, that they stung worse than nettles. Since that time, I have never touched a palmer-worm with my fingers.

It was perhaps a proper punishment for neglecting the knowledge that others had recorded. But I always had rather an aversion to book entomology, and used to work out an insect as far as possible, and *then* see what books said about it. Certainly, although not a very rapid mode of work, yet it was a very sure one, and fixed the knowledge in the mind.

On the same plate, fig. 4 *a*, is shown the caterpillar of this moth, a creature conspicuous from the tufts of beautifully-coloured hair which are set on its body like camel-hair brushes.

The caterpillar spins for itself a silken nest wherein to pass its pupa state, and in general there is nothing remarkable about the nest. But I have one in my collection of insect habitations that is very curious.

I had caught, killed, and pinned out a large dragon-fly, and placed it in a cardboard box for a time. Some days afterwards, a palmer-worm had been captured, and was imprisoned in the same box. I was not aware that such a circumstance had happened, and so did not open the box for a week or two, when I expected to find the dragon-fly quite dry and ready for the cabinet.

When, however, the box was opened, a curious state of matters was disclosed. The caterpillar had not only spun its cocoon, but had shredded up the dragon-fly's wings, and woven them into the substance of its cell. The glittering particles of the wing have a curious effect as they sparkle among the silver fibres.

On plate D, fig. 3 *a*, is represented a creature whose sole claim to admiration is its domestic virtue, for elegance or beauty it has none. It hardly seems possible, but it is the fact, that this clumsy creature is the female Vapourer Moth, the male being represented immediately below fig. 3.

Why the two sexes should be so entirely different in aspect, it is not easy to understand. The female has only the smallest imaginable apologies for wings, and during her whole lifetime never leaves her home, seeming to despise earth as she cannot attain air.

This moth is not obliged to form laboriously a warm habitation for her eggs, for she places them in a silken web which she occupied in her pupal state, and from which she never travels.

Curiously enough, her eggs are not placed within the hollow of the cocoon as might be supposed, but are scattered irregularly and apparently at random over its surface. Even there, though, they are warm enough, for the cocoon itself is generally placed in a sheltered spot, so that the eggs are guarded from the undue influence of the elements, and at the same time protected from too rapid changes of temperature.

In the hot summer months, the leaves of trees are crowded with insects of various kinds, which fly out in alarm when the branches are sharply struck. Oak trees are especially insect-haunted, and mostly by one species of moth, a figure which is given on plate B, fig. 1.

This little moth is a pretty object to the eyes, but a terrible destructive creature when in its caterpillar state, compensating for its diminutive size by its collective numbers. The caterpillar is one of those called "Leaf-Rollers," because they roll up the leaves on which they feed, and take up their habitation within.

There are many kinds of leaf-rollers, each employing a different mode of rolling the leaf, but in all cases the leaf is held in position by the silken threads spun by the caterpillar.

Some use three or four leaves to make one habitation, by binding them together by their edges. Some take a single leaf, and, fastening silken cords to its edges, gradually contract them, until the edges are brought together and there held. Some, not so ambitious in their tastes, content themselves with a portion of a leaf, snipping out the parts that they require and rolling it round.

The insect before us, however, requires an entire leaf for its habitation, and there lies in tolerable security from enemies. There are plenty of birds about the trees, and they know well enough that within the circled leaves little caterpillars reside. But they do not find that they can always make a meal on the caterpillars, and for the following reason.

The curled leaf is like a tube open on both ends, the caterpillar lying snugly in the interior. So, when the bird puts its beak into one end of the tube, the caterpillar tumbles out at the other, and lets itself drop to the distance of some feet, supporting itself by a silken thread that it spins.

The bird finds that its prey has escaped, and not having sufficient inductive reason to trace the silken thread and so find the caterpillar, goes off to try its fortune elsewhere. The danger being over, the caterpillar ascends its silken ladder, and quietly regains possession of its home.

Myriads of these rolled leaves may be found on the oak trees, and the caterpillars may be driven out in numbers by a sharp jar given to a branch. It is quite amusing to see the simultaneous descent of some hundred caterpillars, each swaying in the breeze at the end of the line, and occasionally dropping another foot or so, as if dissatisfied with its position.

Each caterpillar consumes about three or four leaves in the whole of its existence, and literally eats itself out of house and home. But when it has eaten one house, it only has to walk a few steps to find the materials of another, and in a very short time it is newly lodged and boarded.

The perfect insect is called the "Green Oak Moth". The colour of its two upper wings is a bright apple green; and as the creature generally sits with its wings closed over its back, it harmonises so perfectly with the green oak leaves, that even an accustomed eye fails to perceive it. So numerous are these little moths, that their progeny would shortly devastate a forest, were they not subject to the attacks of another insect. This insect is a little fly of a shape something resembling that of a large gnat; and which has, as far as I know, no English name. Its scientific title is Empis. There are several species of this useful fly, one attaining some size; but the one that claims our notice just at present is the little empis, scientifically Empis Tessellata.

I well remember how much I was struck with the discovery that the empis preyed on the little oak moths, and the manner in which they did so.

One summer's day, I was entomologising in a wood, when a curious kind of insect caught my attention. I could make nothing of it, for it was partly green, like a butterfly or moth, and partly glittering like a fly, and had passed out of reach before it could be approached. On walking to the spot whence it had come, I found many of the same creatures flying about, and apparently enjoying themselves very much.

A sweep of the net captured four or five; and then was disclosed the secret. The compound creature was, in fact, a living empis, clasping in its arms the body of an oak moth which it had killed, and into whose body its long beak was driven. I might have caught hundreds if it had been desirable. The grasp of the fly was wonderful, and if the creature had been magnified to the human size, it would have afforded the very type of a remorseless, deadly, unyielding gripe. Never did miser tighter grasp a golden coin, than the empis fastens its hold on its green prey. Never did usurer suck his client more thoroughly than the empis drains the life juices from the victim moth.

He is a terrible fellow, this empis, quiet and insignificant in aspect, with a sober brown coat, slim and genteel legs, and just a modest little tuft on the top of his head. But, woe is me for the gay and very green insect that flies within reach of this estimable individual.

The great hornet that comes rushing by is not half so dangerous, for all his sharp teeth and his terrible sting. The stag-beetle may frighten our green young friend out of his senses by his truculent aspect and gigantic stature. But better a thousand stag-beetles than one little empis. For when once the slim and genteel legs have come on the track of the little moth, it is all over with him. Claw after claw is hooked on him, gradually and surely the clasp tightens, and when once he is hopelessly captured, out comes a horrid long bill, and drains him dry. Poor green little moth!

Still continuing our research among the oak leaves, we shall find many of them marked in a very peculiar manner. A white wavy line meanders about the leaf like the course of a river, and, even as the river, increases in width as it proceeds on its course. This effect is produced by the caterpillar of one of the leaf-mining insects, tiny creatures, which live between the layers of the leaf, and eat their way about it.

Of course, the larger the creature becomes, the more food it eats, the more space it occupies, and the wider is its road; so that, although at its commencement the path is no wider than a needle-scratch, it becomes nearly the fifth of an inch wide at its termination. It is easy to trace the insect, and to find it at the widest extremity of its path, either as caterpillar or chrysalis. Often, though, the creature has escaped, and the empty case is the only relic of its being.

There are many insects which are leaf-miners in their larval state. Very many of them belong to the minutest known examples of the moth tribes, the very humming bird of the moths, and, like the humming birds, resplendent in colours beyond description. These Micro-Lepidoptera, as they are called, are so numerous, that the study of them and their habits has become quite a distinct branch of insect lore.

Some, again, are the larvæ of certain flies, while others are the larvæ of small beetles. Their tastes, too, are very comprehensive, for there are few indigenous plants whose leaves show no sign of the miner's track, and even in the leaves of many imported plants the meandering path may be seen.

There are some plants, such as the eglantine, the dewberry, and others, that are especially the haunts of these insects, and on whose branches nearly every other leaf is marked with the winding path. I have now before me a little branch containing seven leaves, and six of them have been tunnelled, while one leaf has been occupied by two insects, each keeping to his own side.

The course which these creatures pursue is very curious. Sometimes, as in the figure on plate A, fig. 1, the caterpillar makes a decided and bold track, keeping mostly to the central portion of the leaf.

Sometimes it makes a confused tortuous jumble of paths, so that it is not easy to discover any definite course.

Sometimes it prefers the edges of the leaves, and skirts them with strange exactness, adapting its course to every notch, and following the outline as if it were tracing a plan.

This propensity seems to exhibit itself most strongly in the deeply cut leaves. And the shape or direction of the path seems to be as property belonging to this species of the insect which makes it; for there may be tracks of totally distinct forms, and yet the insects producing them are found to belong to the same species.

If the twigs of an ordinary thorn bush be examined during the winter months, many of them will be seen surrounded with curious little objects, called "fairy bracelets" by the vulgar, and by the learned "ova of Clisiocampa Neustria". These are the eggs of the Lackey Moth, and are fastened round the twigs by the mother insect, a brown-coloured moth, that may be found in any number at the right time.

It is wonderful how the shape of the egg is adapted to the peculiar form into which they have to be moulded, and how perfectly they all fit together. Each egg is much wider at the top than at the bottom; and this increase of width is so accurately proportioned, that when the eggs are fitted together round a branch, the circle described by their upper surfaces corresponds precisely with that of the branch.

These eggs are left exposed to every change of the elements, and are frequently actually enveloped in a coat of ice when a frost suddenly succeeds a thaw. But they are guarded from actual contact with ice and snow by a coating of varnish which is laid over them, and which performs the double office of acting as a waterproof garment and of gluing the eggs firmly together. So tightly do they adhere to each other, that if the twig be cut off close to the bracelet the little egg circlet can be slipped off entire.

CHAPTER VIII.

LAPPET MOTH—BRIMSTONE MOTH—ITS CATERPILLAR—CURRANT MOTH—CLEAR-WINGS—WHITE-PLUME MOTH—TWENTY-PLUME MOTH—ADELA—AN INSECT CINDERELLA—NAMING INSECTS—THE ATALANTA—AN INSECT CRIPPLE—PEACOCK BUTTERFLY—BLUE AND OTHER BUTTERFLIES.

LAPPET MOTH.

The accompanying cut is a good representation of a very singular creature called the "Lappet Moth". As may be seen by the engraving, when it is settled quietly upon a leaf with folded wings, it bears a closer resemblance to a bundle of withered leaves than to any living creature. In this strange form lies its chief safety, for there are few eyes sufficiently sharp to detect an insect while hiding its character under so strange a mask.

There are several other examples of this curious resemblance between the animal and vegetable kingdoms, one or two of which will be mentioned in succeeding pages.

The name of "Lappet Moth" is hardly applicable, as it ought rather to be called the moth of the lappet caterpillar. This title is given to the creature because it is furnished with a series of fleshy protuberances along the sides, to which objects the name of lappets has been fancifully given.

It is generally supposed to be a rare moth; but I have not found much difficulty in procuring specimens either in the larval state or as moths. Both moth and caterpillar are of a large size, the caterpillar being about the length and thickness of a man's finger. Its colour is a tolerably dark grey, but subject to some variation in tint. There is no difficulty in ascertaining this species of the creature, as it is clearly distinguished from caterpillars of a similar shape or line by two blue marks on the back of its neck, as if a fine brush filled with blue paint had been twice drawn smartly across it. The curious "lappets" too are so conspicuous that they alone would be sufficient for identification.

One of the examples of animal life simulating vegetation now comes before us in the person of the Brimstone Moth, or rather its caterpillar.

This is a very common insect, and may be recognised at once by its portrait on plate C, fig. 3.

The caterpillar is represented immediately above, fig. 3 *a*. This is one of the caterpillars called "Loopers," on account of their peculiar mode of walking.

They have no legs on the middle portion of their bodies, but only the usual six little legs at the three rings nearest the head, and a few false legs by the tail; so when they want to walk, they attain their object by holding fast with their false or pro-legs as they are called, and stretching themselves forward to their fullest extent. The real legs then take their hold, and the pro-legs are drawn up to them, thus making the creature put up its back like an angry cat.

The grasp of the pro-legs is wonderfully powerful, and in them lies the chief peculiarity of the creature. The surface of the body is of a brownish tint, just resembling that of the little twigs on which it sits; there are rings and lines on its surface that simulate the cracks and irregularities of the bark, and in one or two places it is furnished with sham thorns.

Trusting in its mask, the caterpillar grasps the twig firmly, stretches out its body to its full length, and so remains, rigid and immovable as the twigs themselves. People have been known to frighten themselves very much by taking hold of a caterpillar, thinking it to be a dead branch.

The only precaution taken by the creature is to have a thread ready spun from its mouth to the branch, so that if it should be discovered, it might drop down suddenly, and when the danger was over, climb up its rope and regain its home.

The commonest of the loopers is the well-known caterpillar of the Currant or Magpie Moth, plate E, fig. 3. This creature is remarkable from the circumstance that its colours are of the same character throughout its entire existence; the caterpillar, chrysalis, and perfect moth showing a similar rich colour and variety of tint, as seen on figs. 3 *a* and 3 *b*.

It is a curious fact that almost every stratagem of animals is used by man; whether intuitively, or whether on account of taking a hint, I cannot say.

For example, Parkyns, the Abyssinian traveller, tells an amusing tale of a party of Barea robbers, who when pursued got up a *tableau vivant* at a moment's notice. One man personated a charred tree-stump, and the others converted themselves into blackened logs and stones lying about its base.

It seemed so impossible for human beings to remain so still, that a rifle-ball was sent towards the stump, and caused it to take to its heels, followed by the logs and stones.

I have heard of a similar stratagem that was put in force by a robber who was interrupted on his way into the tent by the appearance of its inmate, an officer. He was so completely deceived, that he actually hung his helmet on one of the branches, which branch was in fact the robber's leg. The joke was almost too good, but the stump stood fast, until the officer leaned his back against it. Officer and stump came to the ground together, and the stump escaped, carrying off the helmet as a trophy. I think that he deserved it.

I conclude this chapter with a short notice of five beautiful and curious little moths.

The first of these, the "Currant Clear-wing," is frequently mistaken for a gnat or a fly, and it is sometimes a difficult task to persuade those who are unaccustomed to insects that it can be a moth. As a general rule, the wings of moths are covered with feathers, and many are even as downy in their texture as the plumage of the owl. But there is a family of moth, called the clear-wings, whose wings are as transparent as those of bees or flies. Some of these are as large as hornets, and resemble these insects closely in general aspect.

Some fourteen or fifteen species of these curious creatures are found in England; and each of them bears so close a likeness to some other insect, that it is named accordingly. For example, the species which we are now examining is called the "gnat-like Egeria," another is the "bee-like," another the "hornet-like," another the "ant-like," and so on. Plate A, fig. 3.

The currant clear-wing may be found on the leaves of currant bushes, where it loves to rest. In 1856 I took a great number of them in one small garden, often finding two or more specimens on one currant bush.

Next come two beautiful examples of the Plume Moth, the White Plume and the Twenty Plume.

The first of these insects is very common on hedges or the skirts of copses, and comes out just about dusk, when it may be easily captured, its white wings making it very conspicuous. See Plate H, fig. 9.

The chief distinguishing point in the plume moth is that the wings are deeply cut from the point almost to the very base, and thus more resemble the wings of birds than those of insects.

In the white plume there are five of these rays or plumes, three belonging to the upper pair of wings and two to the lower.

From the peculiarly long and delicate down with which the body and wings are covered, it is no easy matter to secure the moth without damaging its aspect. The scissors-net is, perhaps, the best that can be used for their capture; for, as they always sit on leaves and grass with their wings extended, they are inclosed at once in a proper position, and cannot struggle. A sharp pinch in the thorax from the forceps, which a collector ought always to have with him, kills the creature instantly; for it holds life on very slender tenure. The slender entomological pin can then be passed through the thorax, while the net is still closed, and thus the head of the pin can be drawn through the meshes of the net when it is opened.

In this way the moth may be preserved without the least injury to its appearance, or without ruffling the vanes of one of its beautiful plumes.

Of all the plume moths this is the largest, as a fine specimen will sometimes measure more than an inch across the wings. There is a brown species, nearly as large, and quite as common; but which is often overlooked on account of its sober colouring; and as often mistaken for a common "daddy-long-legs," to which fly it bears a close resemblance.

The Twenty-plume Moth (plate C, fig. 9) is hardly named as it deserves; for as the wings on each side are divided into twelve plumes, it ought to be named the twenty-four plume. A better title is that of the "Many-plume Moth".

It is very much smaller than either of the preceding "plumes"; and its radiating feathers are so small and so numerous, that at a hasty glance it scarcely seems to present any remarkable structure. It must be examined with the aid of a magnifying glass before its real beauty can be distinguished.

The moth is common enough, and may be easily caught, as it has a strange liking for civilised society, and constantly enters houses. As insects generally do, it flies to the window, and scuds unceasingly up and down the panes of glass, just as if it wished to make itself as conspicuous as possible.

The last of our moths is the beautiful Long-horn, for a figure of which see plate H, fig. 4. Another Long-horn Moth, the Green Adela, is shown on plate C, fig. 10. It is nearly as common as the last-mentioned insect.

It is a horrid name, for its agricultural associations are so potent, that the idea conveyed to the mind by the term "Long-horn" is that of a huge bovine quadruped, with sleek solid sides telling of oil-cake, with horns that are long enough to spike four men at once, two on each horn, and with a ponderous tread that rivals that of the hippopotamus.

Whereas, our little moth is the epitome of every fragile, fairy-like beauty, and seems fitter for fairy tale, "once upon a time," than for this nineteenth century. Its "horns," as the antennæ are called, are wondrously long and slender. I have just taken measurement of one of these moths, and find that the body and head together are barely a quarter of an inch in length, while the antennæ are an inch and a quarter long. It is hardly possible to conceive any living structure more delicately slender than their antennæ. The moth delights in sunny glades, as so sunny a creature ought to do; it sits on a leaf, basking in the glaring sunbeams, while its antennæ, waving about in graceful curves, are only to be traced by the light that sparkles along them. They are as slender as the gossamer threads floating in the air, and like them only seen as lines of light. They are too delicate even for Mab's chariot traces. The grey-coated gnat might use one of them as his whip: but it would only be for show, as beseemeth the whip of a stage-coach; for it could not hurt the tiniest atomy ever harnessed.

And yet the little Adela, for such is her scientific title, flies undauntedly among the trees, threading her way with perfect ease through the thickest foliage, her wondrous antennæ escaping all injury, and gleaming now and then as a stray sunbeam touches them.

There is nothing very striking in the Adela's external appearance; she is just a pretty, unobtrusive, bronze-coloured little thing, from whom many an eye would turn with indifference, if not with contempt. Truly, in vain are there pearls, while the swinish nature prefers dry husks.

Place this quiet, bronze-coloured little creature under a microscope, and Cinderella herself never exhibited such a transformation. The mind of man has never conceived a robe so gorgeous as that which enwraps a small brown moth. Refulgent golden feathers cover its body and wings, sparkling gemlike points scatter light in all directions, while on the edges of each feather rainbow tints dance and quiver. It seems as if the creature wore two robes—a loose golden-feather vesture above, and the rainbow itself beneath. Each fibre of the fringe that edges the wings is a prism, and even the slender antennæ are covered with golden feathers. Words cannot describe the wondrous beauty of this creature.

Methinks a view of these earthly creatures can the better enable one to appreciate the ineffable glories of the heavenly beings. Even the earth-insect is beautiful beyond the power of words to describe—how much more so the heavenly angel!

When the study of entomology first rose to the dignity of a science, it was found necessary that each insect should be distinguished by a definite title. Formerly, it was necessary to describe the insect when speaking of it; and in consequence both cabinets and memories were overloaded with words.

For example, the Meadow-brown Butterfly was named "Papilio media alis superioribus superne media parte rufis". In English: "The middle-sized butterfly, the centre of whose upper wings are reddish on the upper surface". Cromwell's Puritan soldier might have taken a lesson in nomenclature from an entomologist cabinet; and it is not easy to say which would occupy the greater time in reading, the list of butterflies or the regimental roll-call. These difficulties being patent, the nomenclators leaped at once, as is the habit of human nature, into the opposite extreme; and so, instead of making an insect name an elaborate description of its appearance, gave it a title which did not describe it at all, and would have been just as applicable to any other insect. Old Homer's pages afforded a valuable treasury of names; and accordingly, Greek and Trojan may reasonably be astonished to find their names again revived on earth.

Even our British butterflies have appropriated Homeric titles. For example, the two first on the list are named Machaon and Podalirius, known to students of Homer as the two medical officers that accompanied the Greek army.

Numerous, however, as are the Homeric heroes and heroines, the insects far outnumbered them. So, after exhausting Homer, the dramatists were called into requisition, and plundered of their "personæ". Fiction failing, history, or that which is dignified by the name of history, was next sought; and kings, queens, generals, and statesmen lent their names to swell the insect catalogue.

The Latin authors now are required to make up the deficiency, Terence being especially useful. We have in our English list Davus, Pamphilus, and Chrysis, all out of one play, the "Andria".

At last, when Greek and Latin, prose and verse, history and mythology, had been quite exhausted, some enterprising and imaginative men boldly invented new names for new insects. The import of the name was of no consequence to them, and any harmonious combination of syllables was all that they required. Many a valuable hour have they wasted, or rather caused others to waste, in seeking through lexicons and dictionaries for the purpose of discovering the derivation of those unmeaning and underived names.

At last men of science began to see that the name ought to be descriptive of the creature, or its habits, and yet as short as possible; and when this idea was matured, true nomenclature began. In the reformed system, insects are gathered together in societies, through which some general characteristic runs, and each individual bears the name of its genus, as the society is called; and also a second name that distinguishes its species.

The first butterfly which will be mentioned in these pages is seen figured on plate D, fig. 4; and very appropriately bears the name of Atalanta. Those skilled in mythology, or Mangnall's skimmings thereof, will remember that Atalanta was a young lady, so swift of foot that she could run over the sea without splashing her ankles, or on the corn-fields without bending an ear of corn under her weight. The flight of this butterfly is so easy and graceful, that poetical entomologists invested it with the name of the swift-footed Atalanta.

Also it is called the Scarlet-Admiral, in which two names is to be seen the confusion respecting sexes which is found in nautical matters generally. Perhaps the discrepancy might have been avoided by calling the butterfly Cleopatra, that lady being her own admiral.

Few insects are so conspicuous, or have so magnificent an effect on the wing, as the Atalanta; its velvety-black wings, with their scarlet bands, white spots, and azure edges, presenting a bold contrast of colour that is seldom seen, and in its way cannot be surpassed. It is certainly a grand insect; and it seems to be quite aware of its own beauty as it comes sailing through the sunny glades, gracefully inclining from side to side, as if to show its colours to the best advantage. Perhaps its best aspect is when it sits upon a teazle-head, quietly fanning its wings in the sun; for the quiet purple and brown tints of the teazle set off the magnificent pure colours of the insect.

These brilliant colours are only found on the upper surface of the wings, the under surface being covered with elaborate tracery of blacks, browns, ambers, sober blues, and dusky reds, so that when the wings are closed over the creature's back, it is hardly to be distinguished from a dried leaf, unless examined closely.

This distinction of tint often proves to be the insect's best refuge; for, if it can only slip round a tree or a bush, it suddenly settles on some dark spot, shuts up its wings, and there remains motionless until the danger is past. The rough, brown elm bark is a favourite refuge under these circumstances; and it takes a sharp eye to discover the butterfly when settled.

Sometimes the creature is not quite so magnificent, and even appears shorn of its fair proportions. I have now such a specimen before me, which I found on a sandy bank, unable to fly.

My attention was drawn to it by observing a curious fluttering movement of the grasses that covered the bank; and on going up to the spot to see what was the cause, I discovered an Atalanta butterfly that had apparently lost both wings of the left side, and was endeavouring to fly with the remaining pair.

Of course it could only make short leaps into the air, turn over, and again fall to the ground. Wishing to put it out of pain, I killed it, and on examination found that it had never been endowed with wings on its left side, and that those organs had still remained in the undeveloped state in which they had lain under the chrysalis case. Even the right pair had not attained their full development; but in every other respect the insect was perfect.

I suppose that the caterpillar must have selected too dry a spot for its habitation when it became a pupa; and that in consequence the pupa shell was so dry and hard that the butterfly could not make its escape in proper time, I have often seen similar examples in my own caterpillar-breeding experiences. There are also in one of my insect cases two specimens of the little white butterfly, which have met with even a worse fate; for they have not been able to escape at all out of the chrysalis, and so present the curious appearance of a chrysalis furnished with head, antennæ, wings, and legs. The cause of the disaster was probably the same in both cases.

The caterpillar of the Atalanta is shown on plate D, fig. 4 *a*, and is a creature worthy of notice.

It is a well-known saying, that "what is one man's meat, is another's poison"; and the proverb holds good in the case of the Atalanta caterpillar. For its meat is the common stinging-nettle, which is, undoubtedly, poisonous enough to qualify any such proverb.

The colour of the caterpillar is green-black, and along each side runs a spotty yellowish band. Its general shape and appearance can be seen by referring to the figure.

After passing through the usual coat-changing common to all caterpillars, it begins, just before its last change, to prepare a spot where it may pass its pupal state. Its mode of so doing is very curious, and is briefly as follows:—

The chrysalis is intended to remain in an attitude which we should think singularly uncomfortable, but which seems to suit the constitution of certain creatures, such as bats and chrysalides; namely, with its head downward. Why some insects should be thus suspended, while others lie horizontally, is not known as yet. But there can be no doubt but that some purpose is served by the various positions and localities assumed by insects in their pupal state.

Any one of a reflective mind, on hearing that a chrysalis was to be suspended by its tail, would feel some perplexity as to the means by which such a position could be attained. For the old caterpillar's skin has to be shed, and thus the legless, limbless chrysalis is left without any apparent power to suspend itself. The attitude which it assumes may be seen on plate D, fig. 4 *b*. On examining the chrysalis itself, and the leaf or twig to which it is suspended, it will be seen that a little silken mound is fastened to the leaf, and the chrysalis is furnished with some hooked processes on its tail, which are hitched upon the silken threads, and thus hold the creature in the proper position.

The Peacock Butterfly, plate H, fig. 8, is an insect of very similar habits and manners. The under side of the wings is very dark, and when they are closed over the back, the butterfly looks more like a flat piece of brown paper than an insect. The spots on the upper surface of the wing are especially beautiful; and the mode in which those spots are coloured by their feathers is shown in plate L, fig. 4, where a portion of the wing-spot is slightly magnified. This figure shows also the manner in which the feather-dust of the butterfly's wing is arranged. The larva of this beautiful insect is shown on fig. 8 *b*. Like that of the Atalanta, it feeds on the stinging-nettle.

On plate D, fig. 1, is drawn a very lovely insect, one of the numerous blue butterflies that may be seen flitting about the flowers in a garden, themselves of so flower-like an azure, that they may often be mistaken for a blue blossom. The caterpillar, fig. 1 *b*, is, as may be seen, rather curious in shape, and the pupa, fig. 1 *c*, is hardly less so.

Among the scales of this insect occur certain specimens called from their shape "battledore" scales, some of which may be seen on plate K, fig. 8, contrasted with the ordinary scales.

On the same plate as the blue butterfly, fig. 2, is seen a very pretty and common insect, called the "Orange-tip," on account of the colour of the wings. Only the male butterfly possesses these decorations, the female having wings merely white above, although she retains the beautiful green speckling of the under-wings.

Two more butterflies, and those the commonest of all, will complete this chapter. One will be at once recognised from the drawing, plate I, fig. 4, as the White Cabbage Butterfly. The specimen here represented is the female; the male is smaller and has darker spots.

This is the parent of those green and black caterpillars which devastate our cabbage-beds, make sieves of the leaves, and are so disagreeably tenacious of their rights of possession. Pest as it is to the gardeners, to cooks, and sometimes, alas! to consumers, it would be a hundredfold worse but for the exertions of a fly so small as hardly to be noticed, but by its effects. This insect belongs to the same order as the bees, and is shown upon plate J, fig. 6. Small though it be, one such insect can compass the destruction of many a caterpillar, though not one thousandth part of the size of a single victim. While the caterpillar is feeding, the ichneumon fly, as it is called, settles upon its back, pierces its skin with a little drill, wherewith it is furnished, and in the wound deposits an egg. This process is repeated until the ichneumon's work is done.

As each wound is made, the caterpillar seems to wince, but shows no farther sense of uneasiness, and proceeds with its eating as usual. But its food serves very little for its own nourishment, because the ichneumon's eggs are speedily hatched into ichneumon grubs, and consume the fatty portions of the caterpillar as fast as it is formed.

In process of time the caterpillar ought to take the chrysalis shape, and for that purpose leaves its food and seeks a convenient spot for its change.

That change never comes, for the ichneumons have been growing as fast as the caterpillar, with whose development they keep pace. And no sooner has their victim ceased to feed, than they simultaneously eat their way out of the doomed creature, and immediately spin for themselves a number of bright yellow cocoons, among which the dying caterpillar is often hopelessly fixed. Sometimes it has sufficient strength to escape, but it never survives.

In the later summer months, these cocoon masses may be seen abundantly on walls, palings, and similar spots.

Plate I, fig. 3, shows the Brimstone Butterfly, one of the first to appear as the herald of spring.

CHAPTER IX.

STAG-BEETLE—MUSK-BEETLE—TIGER-BEETLE—COCK-TAIL—VARIOUS BURYING-BEETLES—ROSE-BEETLE—GLOW-WORM—GROUND AND SUN-BEETLES, ETC.—HUMBLE-BEES, HORNETS AND THEIR ALLIES—DRAGON-FLIES—CADDIS-FLY—WATER BOATMAN—CUCKOO-SPIT—HOPPERS, EARWIG, AND LACE-FLY.

Of the remaining objects, only a very brief description can be given. Enough, however, will be said to assist the observer in identifying the object, and to serve as a guide to its locality and manners. We will first take the beetles; and as the largest is the most conspicuous, the great Stag-beetle shall have the precedence.

This insect (plate E, fig. 5) is quite unmistakable; and, from its very ferocious aspect, would deter many from touching it. But it is very lamb-like in disposition, and sometimes as playful as a lamb. Its numerous jaws can certainly pinch with much violence; but are not used for the purpose of killing other creatures, as might be supposed.

The food of the stag-beetle is simply the juices of plants, which it sweeps up with that little brush-like organ that may be seen in the very centre of the jaws. In winter it buries itself in the ground, and then, making a smooth vault, abides the winter's cold unharmed.

Only the male beetle possesses these tremendous jaws; those of the female being hardly one-tenth of their size, but so sharp at their points that their bite is just as severe.

The insect that next comes under notice is the Musk-beetle (plate I, fig. 7), a beautiful and conspicuous insect, of a rich green colour above, and a purplish blue below. Its name of musk-beetle is derived from the fragrant scent which it emits; a scent, however, not the least like musk, but more resembling that of roses. It is so powerful that the presence of the insect may often be detected by the nostrils, though it is hidden from the eyes. It may be found chiefly on willow trees.

There is another beetle that gives out a sweet scent, much resembling that of the verbena leaf. This is the Tiger-beetle (plate D, fig. 8). With the exception of the white spots on the wing covers, the colours of this insect are much the same as those of the musk-beetle.

Its name seems hardly commensurate with its aspect; but never was a title better deserved. And, space allowing, I could here draw a terrible character;

but as brevity is enforced, I can but say that this sparkling and beautiful insect seems to have the spirit of twenty tigers compressed into its little body.

All things have their opposites; and opposed to these perfume-bearing beetles are some who are just insect skunks. Chief among these is the common black Cock-tail, a creature of truly diabolical aspect. It is a carrion eater, and intensifies the carrion odour. Still, repulsive as it is, it has its beauties. Its wings are very beautiful, and the mode in which these organs are packed away under their small cases is most wonderful. It is to aid in this process that the cock-tail possesses the faculty of turning its tail over its back. Plate H, fig. 12.

Another beetle of an abominable odour is the Burying-beetle, one of which is shown on plate C, fig. 8. There are many burying-beetles, but this species is the most common.

Their name is derived from their habit of burying any piece of meat or dead animal that may be lying on the surface of the earth, not so much for the sake of themselves as for their progeny. In the buried animal their eggs are laid, and its putrefying substance affords them nourishment. The rapidity with which these and similar insects will consume even a large animal is marvellous. I have seen a large sheep stripped to the very bones in three days, nothing but bones and wool being left to mark the spot where it had lain.

Another kind of burying-beetle is seen on plate B, fig. 7; but instead of dead meat it buries the droppings of living animals, those of the cow being preferred. For this purpose it drives a perpendicular shaft into the ground, makes up a round ball of the droppings, puts an egg into the middle of the ball, rolls it into the hole, and after pushing some earth after it, sets to work at another shaft.

It is evident how beneficial the labours of these insects must be; for by their means the earth is pierced with passages for air—part is thrown out on the surface, where it becomes regenerated by the atmosphere—noxious substances are removed from the surface, where they would do harm, and placed deep in the ground, where they do good.

The popular name for this beetle is the Watchman, because in the dusk of the evening it "wheels its drowsy flight," much as watchmen made their sleepy rounds. It belongs to the same family of insects as the sacred Scarabæus of the Egyptians.

On plate C, fig. 11, is depicted the common Rose-beetle so called because it is an insect of refined habits, and chiefly dwells in the bosom of white roses. Yet it loves earth too, and in pursuance of its mission falls from its rose to earth, and there digs a receptacle for its future progeny. But though in earth, it is not of earth; and, burrow as it may, it returns to its rose without a stain upon its burnished wings.

The curious Glow-worm, as it is called falsely, it being a beetle, and not a worm, is shown on plate J, fig. 1. Both the male and female insect give out this light, as I have often seen, though that of the female is the more powerful. The two sexes are very different in appearance, as may be seen by reference to the plate, fig. 1 being the male, and 1 *a* the female. The object of the light is by no means certain, nor the mode in which it is produced.

On the same plate, fig. 11, is seen the Oil-beetle, an eccentric kind of insect, which, when frightened, pours a drop of oil out of every joint, just as if it were a walking oil-barrel with self-acting taps.

One of the commonest beetles, the Ground-beetle, is seen on the same plate, fig. 10. There are very many ground-beetles, but this is one of the handsomest and most conspicuous. The embossment of its upper surface is worth a close examination, and its colouring is peculiarly rich and deep.

Hot sunny days always seem to bring out a host of insects, among which the Sun-beetles are notable examples. One of these insects is shown on plate D, fig. 6. They are beautifully brilliant as they run among the gravel-stones or over paths, their smooth surface glittering in the sun resplendently.

As an aquatic balance to the terrene Sun-beetles, the Whirligigs (plate F, fig. 4) make their appearance on the surface of the water on any light sunny day. What rule they observe in their mazy dance is more difficult to comprehend than the "Lancers" or a cotillon: but that there must be a rule is clear from the wonderful way in which they avoid striking against each other in their passage.

Every one knows the Lady Bird, with its pretty red wings and black spots. Its larva (plate B, fig. 8) is a very singular creature, and destructive withal, spearing and eating Aphides as ruthlessly as Polyphemus impaled and devoured the captured sailors. It has a curious history, but there is no room for it here.

On plate H, fig. 7, is represented one of the many Skipjack-beetles, who afford such amusement to juveniles by their sudden leaps into the air when laid on their backs. This feat is performed by means of a sudden blow of the head and thorax. Farmers, however, are not all amused by it, for it is the parent of the terrible "wire-worm," so deadly a foe to corn and potatoes.

Some insects prefer corn when placed in granaries, and these are the Weevils, whose grubs populate sea-biscuit, and run races across plates for wagers. Nuts also fall victims to the weevil represented on plate I, fig. 9, or rather to its grub, "Time out of mind the fairies' coachmaker".

There is a very common little green weevil shown on plate C, fig. 7, which, although ordinary enough to the unassisted eye, yet under the microscope glows with jewels and gold. It is, in truth, the British Diamond-beetle. An idea of its appearance may be obtained from plate L, fig. 6, but to give the real glory of the colouring is impossible.

One of the little insects called Death-watches is shown on plate J, fig. 8. There are many insects that go by this name, because they make a slight tapping sound with their heads, probably to call their mates; and which sound has been thought to prognosticate death rather than marriage.

The curious Tortoise-beetle is depicted on plate C, fig. 6. Its chief peculiarity is in its larval state, when it carries a kind of parasol, formed from the remains of the leaves on which it has been feeding.

Last and least of the beetles comes one as destructive as it is small, the Turnip-hopper. This little animal, no larger than a small pin's head, does great damage to the turnip crops, and is therefore hated by farmers. It is shown, much magnified, on plate J, fig. 13.

From the beetles we proceed to the Bee tribe; and first take the common Humble-bee, several of which are shown on plate H, fig. 10, representing the "Red-hipped Humble-bee," which mostly makes its nest among stone-heaps. Fig. 11 is the common Humble-bee, that burrows in the ground, and there builds its thimble-like cells. These cells are very irregular in shape, and are affixed to each other without any definite order. Of these two insects, the latter is harmless enough; but the former becomes very fierce if its nest is approached too closely.

A magnified view of some hairs of the Humble-bee is given on plate K, fig. 11.

There are some bees which make their nests in old walls, where they either dig for themselves a hole, or oftener take advantage of a nail-hole, and so save themselves much trouble. One of these bees is shown on plate H, fig. 2, and is chiefly remarkable for the beautifully tufted extremities of its middle pair of legs.

On plate D, fig. 7, is seen the common Hornet, one of the really terrible of our insects. It mostly makes its nest in hollow trees, and it behoves one to keep very clear of the neighbourhood. The nest is made of wood-fibre, nibbled, and made into a primitive papier-maché.

Two of the Saw-flies may be seen on plate J. Fig. 2 is the common green Saw-fly, and fig. 3 the dreaded Turnip-fly. These are called Saw-flies because they are furnished with saw-like implements, by means of which they cut grooves in certain plants, and in those grooves lay their eggs.

Mention has already been made of the little Ichneumon fly. One of these insects is shown magnified on plate J, fig. 12 *a*, and one of the large species is depicted on plate H, fig. 3. The threefold appendage to the tail is the ovipositor, or instrument by means of which they pierce their victims and deposit their eggs.

There are some allied insects that pierce vegetables instead of insects; and one of their works may be seen figured on plate A, where a bramble-branch has been perforated by them. The well-known oak-apples, plate B, fig. 6, are caused by a Cynips, as the little creature is called; and so is the common Bedeguar of the rose, seen on plate C, fig. 2.

The last of these insects that will be named is the beautiful Fire-tail, plate D, fig. 5, one of the most brilliant insects that our island can boast. There are many British species of this insect, but they all much resemble each other, and are insect cuckoos, laying their eggs in the nests of other insects.

From the bees, we pass to the Flies; and first take a most singular insect, shown on plate H, fig. 5. This insect is found on the blackberry blossoms, and the upper part of its body is so transparent that the leaf on which it sits can be seen through it. It is swift of wing and wary, requiring a quick eye and hand for its capture.

On the same plate, fig. 6, is shown one of the traveller's pests, a fly that bites, or rather bores, the skin, and that with such virulence that it can even strike its poisoned dart through a cloth coat, and make its victim to lament for many an hour after.

One of the various hoverer-flies is shown on plate J, fig. 9. The larva of this insect is very remarkable, on account of its curious breathing apparatus. The larva is properly called the Rat-tailed Maggot, and is shown on the same plate, fig. 8 *a*. The body of the creature is found buried in the mud at the bottom of stagnant pools or cisterns, and the respiration is carried on through the telescopic tail, which is long enough to protrude through the mud, and to convey the necessary oxygen to the system through two flexible air-tubes that pass through the "tail".

It will be remembered that in mentioning the Green Oak Moth, the Destroying Empis was also noticed. One of these flies is shown on plate J, fig. 5, with the poor Tortrix in its grasp. Plate K, fig. 1, shows its foot, and fig. 3 its head, together with its long beak.

The beak of this fly somewhat reminds one of the corresponding portion of the Gnat, which insect is not itself depicted, though on plate F, fig. 10, is shown the wonderful little egg-boat which it makes. This insect glues together its eggs in such a manner that they are formed into a true lifeboat, which cannot be upset, or sunk, or filled with water, but floats securely on the surface until the young are hatched. That object accomplished, the gnat-larvæ tumble into the water, and there undergo their transformation.

The last of the two-winged flies that will be mentioned is the common Daddy Long Legs, or Crane-fly, which seems to set such little value on its limbs. It is a very injurious insect in its larval state, feeding on roots, and doing great damage. Plate H, fig. 1, shows a very pretty species, covered with yellow rings.

Every one must have noticed the beautiful and active insects that are with great truth called Dragon-flies. Their habits and peculiarities would demand a volume; and here they can but be mentioned. Plate F, fig. 6, shows the common Flat Dragon-fly, that may be seen chasing and following flies of all sizes, and even butterflies. Fig. 8 is the elegant Demoiselle, the male of which is shown here, with its dark purple spots on the wings and dark blue body. The female is of a uniform green. Its larva is shown at fig. 8 *a*, where the singular leafy gills may be seen at the end of its tail. Fig. 7 shows another very common Dragon-fly, very thin and ringed with blue circlets.

On the same plate, fig. 12, may be seen several varieties of the objects known to fishermen as "Caddis" cases. These are residences built by the larva of the common Caddis, or Stone-fly, which is represented on the same plate, fig. 9.

Still keeping to plate F, and referring to fig. 1, is seen the horrid-looking Water-scorpion, a creature which, though it does not sting, has much of the scorpion nature, and so bites. Fig. 1 *a* shows the same insect as it appears when flying.

At fig. 3 is seen the Water Boatman, so called because it lies on its back, which is ridged like the keel of a boat, and then rows itself about by means of its middle pair of legs, which closely resemble oars.

Fig. 5 shows a very curious object which is common enough on the margin of pools, and runs on the surface of the water as if it were dry land. When alarmed, it shuts up all its legs, and looks just like a piece of dry grass or thin stick.

Another insect much resembling it, is the common Gerris, seen on plate I, fig. 6. It may be seen on every pond or still water, running over its surface, and is furnished with wings wherewith it can fly to great distances. I have found specimens on the tops of hills, far from any water, and hiding under stones out of the sun's heat. Fig. 1 shows the common May-fly.

All gardeners have been annoyed with the curious production called the Cuckoo-spit. This proceeds from the larva of one of the hoppers, and on removing the frothy substance, the little soft, greenish insect may be found within. The perfect insect is shown on plate C, fig. 1 a, and the exudation itself at fig. 1.

There is another hopper seen on plate B, fig. 2, called from its colour the Scarlet Hopper. It is common enough on ferns, and may be found chiefly in the open spots of forests where ferns abound.

On plate J, fig. 7 a, is the common Green Grasshopper, as it appears when standing; and on fig. 7, the same insect as it appears when using its wings.

The common Earwig, plate I, fig. 8, is introduced for the purpose of showing the very beautiful wing which this insect possesses, and which is seen expanded at fig. 8 a.

The very lovely, though ill-odoured, Lace-wing Fly is shown on plate J, fig. 4, and its very remarkable eggs at 4 a. Each egg is placed at the end of a footstalk, whereby it is kept out of the reach of certain predacious insects.

Various shells are drawn on one or two of the plates, but there is not space for any description. Their names may be found on the Index to Plates. Plate G contains certain fungi and mosses. Fig. 1 is that peculiar plant which reindeer scrape from under the snow in the winter time. Fig. 2 was once dreaded by rustics as "Witch's butter". Fig. 6 shows the curious Earth-star, chiefly remarkable for its resemblance to the marine Star-fish.

INDEX TO PLATES.

- A. (*Front*)
- 1. Tubercled Gall on Bramble-stem.
- 2. Track of Leaf-Miner on Bramble-leaf.
- 3. Gnat-Clearwing Moth.
- 4. Buff-tip Moth.
- —*a*. Caterpillar of do.
- 5. Privet Moth.
- —*a*. Caterpillar of do.
- 6. Snail (*Helix nemoralis*).
- 7. Do. (*Helix nemoralis*) var.
- 8. Do. (*Helix cantiana*).
- 9. Do. (*Helix ericetorum*).
- 10. Do. (*Helix lapicida*).
- 11. Shell (*Cyclostoma*).
- 12. Do. (*Zonites*).
- 13. Do. (*Helix caperata*).
- 14. Do. (*Pupa*).
- 15. Do. (*Clausilia*).
- B.
- 1. Green Oak Moth (*Tortrix*).
- 2. Scarlet Hopper (*Cercopis*).
- 3. Burnet Moth.
- —*a*. Cocoon of do.
- 4. Puss Moth.
- —*a*. Caterpillar of do.
- 5. Tiger-Moth (Arctia).

- —*a*. Caterpillar of do.
- —*b*. Cocoon of do.
- 6. Oak-galls.
- 7. Watchman Beetle (*Geotrupes*).
- 8. Lady-bird (*Coccinella*).
- —*a*. Larva of do.
- C.
- 1. Cuckoo-spit.
- —*a*. Cuckoo Hopper (*Tettigonia*).
- 2. Bedeguar of Rose.
- 3. Brimstone Moth.
- —*a*. Caterpillar of do.
- 4. Emperor Moth.
- —*a*. Caterpillar of do.
- —*b*. Cocoon of do.
- 5. Elephant Hawk-Moth.
- —*a*. Caterpillar of do.
- 6. Tortoise Beetle (*Cassida*).
- 7. Green Weevil.
- 8. Burying-Beetle (*Necrophorus*).
- 9. Twenty-Plume Moth.
- 10. Green Adela.
- 11. Rose-Beetle.
- D.
- 1. Blue Butterfly (*Alexis*).
- —*a*. Do. Wings closed.
- —*b*. Caterpillar of do.
- —*c*. Pupa of do.

- 2. Orange-tip Butterfly.
- 3. Vapourer Moth, Male.
- —*a.* Do Female.
- 4. Red Admiral.
- —*a.* Caterpillar of do.
- —*b.* Pupa of do.
- 5. Fire-tail (*Chrysis*).
- 6. Sun Beetle.
- 7. Hornet.
- 8. Tiger Beetle.
- —*a.* Do Flying.
- E.
- 1. Drinker Moth, Male.
- —*a.* Do. Female.
- —*b.* Do. Caterpillar.
- —*c.* Do. Cocoon.
- —*d.* Do. Chrysalis.
- —*e.* Do. Eggs.
- 2. Humble-bee Fly (*Bombyllus*).
- 3. Magpie Moth.
- —*a.* Do. Chrysalis.
- —*b.* Do. Caterpillar.
- 4 Gold-tailed Moth (*Porthesia*).
- —*a.* Do. Caterpillar.
- 5. Stag-Beetle.
- F.
- 1. Water Scorpion.
- —*a.* Do. Flying.

- 2. Amber Shell (*Succinea*).
- 3. Water Boatman.
- 4. Whirligig Beetle.
- 5. Hydrometra.
- 6. Dragon-Fly (*Libellula*).
- 7. Do. (*Agrion*).
- 8. Do. Demoiselle (*Calepteryus*).
- —*a.* Do. Larva.
- 9. Stone-Fly (*Phryganea*).
- 10. Eggs of Gnat.
- 11. Caddis-cases, composed—
- *a.* Of flat stones.
- *b.* Of bark.
- *c.* Of sand.
- *d.* Of grass.
- *e.* Of grass-stems.
- *f.* Of shells.
- 12. Water shells.
- *g.* Planorbis.
- *h.* Ancylus.
- *i.* Lymnæus.
- *k.* Paludina.
- G.
- 1. Reindeer Moss (*Cladonia*).
- 2. Witch-butter (*Tremella*).
- 3. Polytrichum.
- 4. Bird-nest Moss (*Nidularia*).
- 5. Xylaria.

- 6. Earth-star (*Geastrum*).
- —*a*. Do. closed.
- 7. Arscyria.
- 8. Cup-moss (*Cenomyce*).
- 9. Scarlet Cup-moss (*Peziza*).
- 10. Marchantia.
- H.
- 1. Crane-fly.
- 2. Mason Bee (*Megachile*).
- 3. Ichneumon (*Pimpla*).
- 4. Adela Long-horn.
- 5. Volucella.
- 6. Sting-fly (*Chrysops*).
- 7. Skipjack Beetle (*Elater*).
- 8. Peacock Butterfly.
- —*a*. Do. wings closed.
- —*b*. Do. Caterpillar.
- 9. White-Plume Moth.
- 10. Red-tailed Humble-Bee.
- 11. Common do.
- 12. Cock-tail Beetle (*Goërius*).
- I.
- 1. May-fly (*Ephemera*).
- 2. Scorpion-fly.
- 3. Brimstone Butterfly.
- 4. Cabbage White Butterfly.
- —*a*. Do. Caterpillar.
- 5. Oak Egger-Moth, female.

- —*a*. Do. Cocoon.
- 6. Gerris.
- 7. Musk Beetle.
- 8. Earwig.
- —*a*. Do. flying.
- 9. Nut Weevil.
- J.
- 1*a*. Glow-worm, male.
- —*b*. Do. female.
- 2. Green Saw-fly (*Tenthredo*).
- 3. Turnip-fly.
- 4. Lace-wing Fly.
- —*a*. Eggs of do. on lilac branch.
- 5. Empis.
- —*a*. Do. killing Oak-moth.
- 6. Ichneumon (*Microgaster*) and cocoons.
- 7. Grasshopper, flying.
- —*a*. Do. walking.
- 8. Death-watch (*Anobium*).
- 9. Hoverer-fly.
- —*a*. Rat-tailed Maggot.
- 10. Ground Beetle (*Carabus*).
- 11. Oil Beetle.
- 12. Cocoon of Microgaster, magnified.
- —*a*. Microgaster, magnified.
- 13. Turnip-hopper (*Haltica*), magnified.
- —*a*. Do. natural size.
- 14. Cyclops, magnified, showing egg-sacs.

- 15. Scarlet Spider (*Trombidium*), magnified.
- —*a.* Do. natural size.
- K. MICROSCOPICAL.
- 1. Foot of Empis.
- 2. Pollen—*a.* Sunflower.
- *b.* Passion Flower.
- *c.* Lily.
- 3. Head of Empis.
- 4. Foot of Male Water-Beetle (*Dyticus*).
- 5. Trunk of Blue-bottle Fly.
- 6. Foot of Frog, showing circulation.
- 7. Petal of Geranium, showing stomata.
- 8. Battledore Scales of Blue Butterfly.
- 9. Scale of Fritillary Butterfly.
- 10. Eye of Butterfly.
- 11. Hairs of Humble-Bee.
- L. MICROSCOPICAL.
- 1 and 3. Scales of various Butterflies.
- 2. Eye of Hemerobius.
- 4. Wing of Peacock Butterfly.
- 5. Poppy seeds.
- 6. Wing-case of Green Weevil.
- 7. Egg of Red Underwing Moth.
- 8. —— of Small White Butterfly.
- 9. —— of Tortoiseshell Butterfly.
- 10. —— of Lathonia Butterfly.

B

C

D

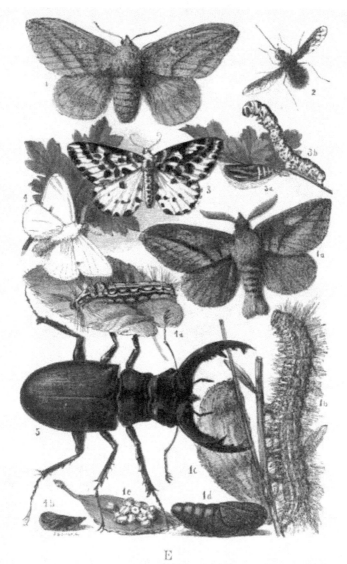

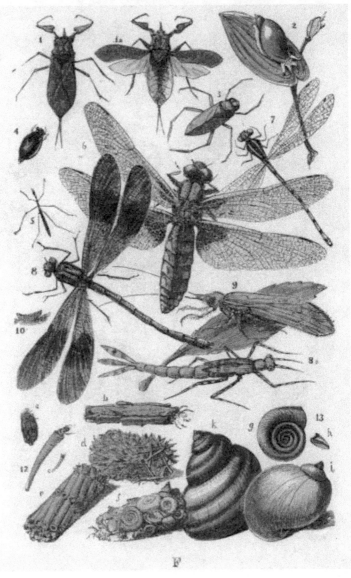

F

II

H

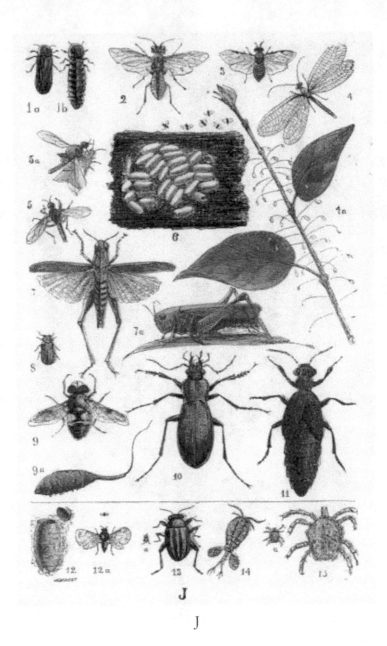

J

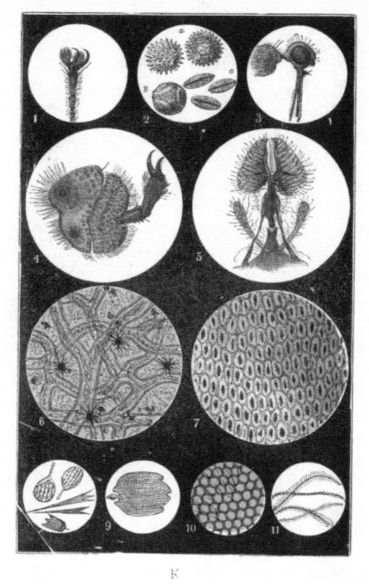

K

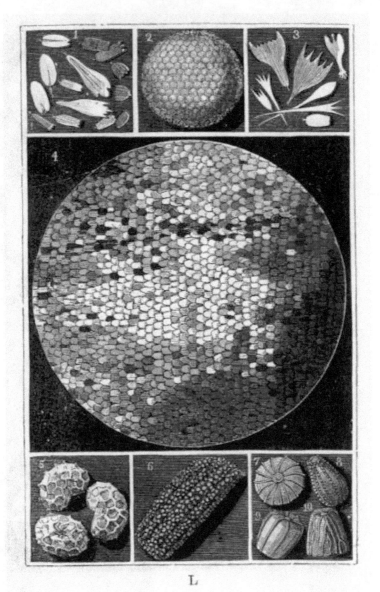

L

CPSIA information can be obtained
at www.ICGtesting.com
Printed in the USA
BVHW031517100122
625871BV00007B/456